I WISH I HAD THOSE GASOLINE PELLETS

I WISH I HAD THOSE GASOLINE PELLETS

CARL KEGERREIS

LitPrime
"Your story is our priority"

LitPrime Solutions
21250 Hawthorne Blvd
Suite 500, Torrance, CA 90503
www.litprime.com
Phone: 1-800-981-9893

Published by LitPrime Solutions   01/18/2023

ISBN: 979-8-88703-118-7(sc)
ISBN: 979-8-88703-119-4(hc)
ISBN: 979-8-88703-120-0(e)

Library of Congress Control Number: 2022923743

CONTENTS

PREFACE

T ODAY, APRIL 25, 2020, I had to take mail to the local post office and I noticed the different prices of gasoline. One sign showed $1.19 per gallon, another $1.47, and the last sign at a local grocery store for $1.32. I was thinking it would be wonderful if the story you are about to read was true, but *I Wish I Had Those Gasoline Pellets* is fiction.

I am a veteran who was drafted into the United States Army in 1961. I was a college junior at Ball State University, Muncie, Indiana, and was short on class credit hours, so I got drafted into the military. I had finished military police school at Fort Gordon, Georgia, and was transferred to Lackland Air Force Base, San Antonio, Texas, to train with military sentry dogs for security at US missile bases. While waiting in my uniform during a layover at the airport in Dallas, Texas, I was approached by a group of people protesting the military. One person spit on my uniform. If my sentry dog, Kim, had been with me, the protesters would never have dared to approach me like that.

I went to the men's room and cleaned the spit from my uniform. Now, years later, while being ordered by our Ohio governor to stay at home due to the pandemic, that memory and the gasoline prices have combined to give you this story to read.

I retired on December 31, 1999, from the CSX Transportation Railroad Federal Police after thirty-three years. I was promoted to lieutenant, captain, and division chief. I have worked cases with the

United States Secret Service, Federal Drug Administration, and Federal Bureau of Investigation in my training. I have also worked with state, county, and local law enforcement agencies. I have been president of the Pennsylvania, Ohio, Indiana, and Michigan Railroad Police Associations. I belonged and supported the Ohio Chiefs of Police with the County Sheriff's Associations in Shelby, Seneca, Wood, and Lucas counties. I have been married to the love of my life for fifty-eight years, and I have three children and five grandchildren.

Like me, you have no doubt noticed how gasoline prices continue to change and how the states continue to add taxes to the price. The prices seem to rise even while you are pumping the liquid into your tank.

You are probably wondering why I decided to write this story about United States Air Force Captain Oley Washington Jr., who was to receive the United States Congressional Medal of Honor from the United States president. To tell the truth, I'm not sure where everything in this story came from—only that I have a colorful imagination. As you read this story, you are going to discover surprises, sadness, danger, exciting adventure, and a new, startling discovery that would be wonderful if it was real. It could save all of us from buying gasoline ever again. Unfortunately, this story is fiction. Sorry about that. But I hope you enjoy it just the same!

Carl Kegerreis with his sentry dog, Kim.

FIGHTING THE ENEMY

U NITED STATES AIR FORCE Captain Oley Washington Jr., with his crew of five in other jet planes, was flying into Vietnam to unleash bombs, bullets, and rockets upon trucks loaded with enemy soldiers. They were moving to the front line to fight our American soldiers.

Oley watched as the bombs they dropped erupted into explosions below and delivered a bomb himself that blew up a truck loaded with the enemy. He moved away from the area, gaining altitude, when a warning light appeared on his radar screen. An enemy rocket was about to hit his plane's tail section. Oley attempted to maneuver his plane but felt the explosion. He was losing control and the plane was falling. He struggled with the controls, but to no avail. Giving up, he quickly pulled the emergency lever, which propelled him out of the plane. His seat fell away and his parachute opened. His crew saw that he had ejected from the plane and protected him as he floated downward. Hitting the ground hard, Oley rolled over covered in his parachute. He dropped the final parachute strap and began crawling toward the woods, moaning from a sharp pain in his leg. The cold air numbed his face. Crawling in the dark thicket, he moved closer to the trees. Animal noises were close and loud, and he told himself, "I'm in a jungle." Finding a fallen old tree covered in brush, he crawled under the branches. At the same time, he could hear and see enemy soldiers running through the jungle near him. He pulled his military .45-caliber pistol loaded with a silencer from his holster and pointed it at them as they ran past him. Satisfied that they didn't see him, he replaced it in his holster. That's when he heard someone shout and saw one of the enemy soldiers holding up his parachute.

The soldiers quickly searched through the jungle while Oley did his best to stay as quiet as possible. He nearly cried out when one of the men hit the brush right below his feet with a stick before being ordered away. After several minutes of waiting, Oley stopped hearing voices. He thanked God that the enemy never found him and sat up. The spinning rotors from a helicopter sounded, and Oley removed a small military radio from his front jacket pocket. He pushed the button, but no red light came on. He could hear the helicopter getting closer. He returned the radio to his pocket and removed his right military boot, massaging his leg. He pulled a small first-aid box from a large pocket in his pants. He then heard the helicopter above him and grabbed the radio from his pocket, pushed the button. Still no red light. He examined the radio, finding a large crack on the back side. He removed the back of the radio, seeing a broken wire. He tapped that one wire and pushed the button, and there was the red light. Oley slowly whispered in the radio, "Angel One, Angel One for the rescue." He waited several minutes, then said, "Angel One, Angel One for the rescue."

He waited, then heard, "Angel One, Angel One, we don't see you. Show yourself."

Oley replied, "Angel One, in woods hiding, leg hurt from landing, and avoiding capture. Land at side of woods with clearing where I can crawl to you." Oley wanted to jump for joy when he heard, "Angel One, see clearing and landing."

He was crawling out of the brush when he heard an explosion and gunshots. He raised up and saw the busted up helicopter smoking and the enemy still shooting their guns at it and its crew. He realized that his wanting to be rescued may have caused their deaths. He lay under the brush and tree when he heard a small noise. He grabbed his .45-caliber revolver, pointing it toward the noise. He couldn't believe what he was seeing: a huge snake was slithering toward him. Oley lay very still, watching the snake. It had a big head, diamond-shaped eyes, and a black tongue moving in and out of its mouth. The snake was moving fast, closer to him, when he fired his gun with a small *pop* sound. The snake no longer moved. He pulled the dead snake to him with a tree branch before sitting up under the brush. He cut the snake up into small portions, thanking God for the food, and ate some raw snake meat and placed the rest in his pockets.

It was getting dark. His plan was to check the busted-up helicopter and the crew at night, hoping to find anyone still alive and needing help. Looking at his watch with a small light, he saw it was now midnight. Satisfied the enemy had left the area, he crawled out of the brush and from under the tree to the helicopter. The helicopter was a mess of steel still smoking from the fire after it crashed. Crawling closer to the helicopter, he listened for the enemy and reached the helicopter's cockpit door. Slowly lifting his body, he looked through the broken window and could see the look of death on the pilot and copilot, still harnessed in their seats. Working his way through the twisted hot steel, he gained access into the cockpit, checking all the officers, but confirming all were dead. Inside, he removed the dog tags on each officer to give to their families. Oley continued moving around the helicopter, finding food to eat. The raw snake meat he ate earlier was causing him severe stomach pain. He ate some crackers and drank raw tomato soup from a can. Looking at his watch, he saw the time was now 2:00 a.m., so he decided to return to his hiding place.

He slowly dropped out of the helicopter and crawled quickly toward the woods. He stopped crawling when he heard the enemy hollering and running toward him. He grabbed his revolver from his holster, which dropped from his hand when he was hit in the head by a rifle. Then he was picked up by enemy soldiers and placed on a dirty military truck floor. Enemy soldiers sat around him hollering, laughing, and talking while pointing their guns at him. Oley closed his eyes and prayed, "God, please forgive me." Oley believed that he soon would be dead like the helicopter crew.

SURVIVAL

SEVERAL YEARS LATER, RED, a homeless man, was behind the Trident Grocery in Detroit, Michigan, climbing into a trash dumpster belonging to the store. Red looked like he was in his seventies, but was really in his sixties. He had long redish-gray hair and a long beard and wore a dirty, ragged shirt, jacket, and pants. He kicked the trash, looking for any special prize that might be hiding in the debris. Turning over a big box, he found a cigar box with a smashed cigar still inside. He reached in the torn pocket on his jacket, attempting to find a match to light the cigar but found none. He chewed on the cigar, spitting out the juice while kicking and turning over the trash to find something to eat or use later. The store door opened on the rear deck, and hearing it, Red climbed out of the dumpster. The store owner, Mike Trident, hollered, "Red, is that you in our trash dumpster?"

Red replied, "Yep, Mr. Trident, it's me."

Mike, a big man, walked down the deck steps carrying a blood-stained wrapped package. Smiling at Red, Mike handed him the package containing pieces of raw meat. Mike shook Red's dirty hand, and with a very concerned look, said, "Red, you have been in my dumpster several years."

Red replied, "Thank ya, Mr. Trident."

Mike removed his hand from Red's, telling him, "I'm afraid one of these days I am going to find your stiff body in our dumpster if you keep getting in it."

Red reached in his front pants pocket, handing Mike several green pellets that looked like jellybeans to Mike.

"Thank ya, Mr. Trident," Red said.

Mike tossed the jellybeans into the parking lot. "Thanks but no thanks, Red," he said. "I don't eat things from a dumpster—and neither should you."

Hearing thunder, Red waved goodbye to Mike and moved away. Mike looked up at the dark clouds and returned to the store.Red opened a gate and was running toward the woods when several neighborhood kids, laughing and hollering, threw balloons loaded with water at him. A few of the balloons hit Red just before he disappeared into the woods.

"Since you never take a bath, we decided to help you, old man!" one of the boys yelled. Red slowed down, stepping over brush until he reached an old vacant building that had been a bank many years before. Red was grateful to have this old brick building as his home.

No one except Red had been in the building for years. It was covered with trees and brush, and Red had created a secret path into the building. He found the place several years ago after he jumped from a an empty freight train box car while it was slowly passing through the woods.

Grabbing some old paper, he dried himself where the balloons had hit him. He found his old frying pan that he'd cleaned earlier with rainwater stored in a wooden barrel. He unwrapped the meat scraps and placed them in the pan on a small gas burner. Reaching up to some old broken shelves above the water barrel, he grabbed a can of beans he had found in Trident's dumpster.

"This'll be a great meal tonight," he muttered to himself. He reached into a busted wooden drawer that held a can opener and several shriveled onions. Opening the can of beans and finding one good onion, he placed them in the pan with the meat scraps.

He wondered whether the small gas stove had enough fuel to cook his food, and was thinking about when he found the gas cylinder in the dumpster behind Wesswell's Gas Company. There was gas when he opened the nozzle, believing the cylinder was about a third full. He didn't ever want to meet the company employee who had threatened him again. The man was big, shouting and cursing while Red dragged the cylinder away from the dumpster. Red offered the angry man some pellets from his pocket, but the man had shouted, cussed, and threw the pellets on the ground, trying to crush them with his foot.

Enough gas was in the cylinder, and he enjoyed his meal of beans, meat scraps, and onion. He cleaned the pan and a fork with rainwater that he collected in a barrel from the roof's runoff. Then he hung the pan and fork on rusty nails that extended from a broken shelf.

In the meantime, Red's water-balloon assailants were caught by Father Pario, a priest from the St. Reba Catholic Church. Father Pario saw the kids throwing water balloons at Red and had approached them after Red ran away. Their parents were members of his church and he recognized them immediately.

"I'll give you kids a choice," he told them sternly. "You can either apologize to that poor old man or face your parents."

The kids agreed to apologize. Father Pario opened the gate, and they walked down the path to the woods ending at huge trees, brush, and high grass. Father Pario called out for the old man. Then the boys hollered, "Hey, old man, we're sorry!"

There was no answer, and the old man didn't show himself.

Father Pario looked down at the boys. "You tried to apologize, and you must now promise me you won't do it again. Do you promise?"

The boys nodded their heads.

"We're sorry," one of the boys said. "Please don't tell our parents, Father." Father Pario looked at them, satisfied. "I won't tell your parents this time, but you must never repeat this. Is that understood?" They nodded their heads and apologized again.

A TRIP TO THE CASINO
AND A SURPRISE

HAYWOOD CARRIED THE MORNING paper into the kitchen. His wife, Lilly, was placing dishes in their dishwasher. Haywood listened to a TV news reporter talk about the strange appearance of gasoline puddles in Detroit, Michigan. "The price of gasoline is now $340 a barrel and rising. Regular gasoline is expected to cost the consumer now $3.50 a gallon with new state taxes, and the price per gallon may climb even higher. Can you believe that gasoline was found all over the Trident Grocery parking lot in Detroit and on the ground at Wesswell's Company close to their rear dumpster? Owner Mike Trident at Trident's Grocery said no one had reported losing any gasoline in his parking lot, and he has no information how the gasoline was placed there. The local fire department was called, and they used foam to remove the gasoline at both locations."

Haywood couldn't believe what he just heard.

He also saw an ad for a great deal at a Casino in Canada and showed it to Lilly. "We should go to Canada," he said. "This could be a fun trip, and we get a free lunch buffet at the casino. The bus fare is only twenty-five dollars."

Lilly looked at Haywood. "How much money are we going to spend at the casino?"

Haywood was not surprised because Lilly handled the family budget. "Sweetheart," he said, "I have an idea. How about we each take only fifty dollars to spend at the casino, and whoever brings home the most money gets another fifty dollars from our budget?"

Lilly smiled. "Can I get the fifty now instead of going to the casino?"

"We'll have a fun day and not spend money on gas," Haywood said, smiling back.

Lilly laughed. "Yes. We'll spend our money at the casino."

Haywood called the local bus terminal, making reservations for himself and Lilly. The next day Haywood and Lilly Runyan boarded the bus by the city mall with the other senior passengers headed for Canada and the casino. The bus seats did not allow much room for Haywood's legs. Lilly was sitting by the bus window after placing a cooler with several bottles of cold water on the floor by her feet. Lilly then placed her head on Haywood's shoulder and closed her eyes. He looked at his lovely wife, his best friend and lover, who he'd been married to for forty years. He then placed his head on her head and closed his eyes as the bus continued moving.

The bus arrived at the Canada casino, and Haywood couldn't believe how fast they crossed over the border from the United States. The bus had stopped at the Canadian gate, and he heard the bus driver tell the customs officer, "I have my bus loaded with seniors going to the casino." The customs officer raised the gate and motioned for the bus to proceed. Haywood wondered why he needed his passport, Lilly's driver's license, and her birth certificate.

The bus driver was middle aged and short with a medium build and a mustache that curled upward to each side of his nose. He was wearing a jacket with a veteran's prisoner of war symbol. The bus driver stood, speaking to the seniors. "People, I need your attention now! This is bus number 218, and I will arrive back here at 4:30 p.m. and will depart from here at 4:45 p.m. You seniors must be on this bus by 4:45 p.m. or you have to find another way home. All you seniors set your watches to my time now, 11:20 a.m. This bus will depart at 4:45 p.m., and you must keep your same seat on this bus when you return. I will not separate any seniors fighting over seats. I now will pass out casino cards you need to fill out before leaving this bus to go in. Do you have any questions?"

Haywood didn't like the bus driver's attitude or having to fill out cards for the casino. However, the cards were required for the casino certificates for the free lunch buffet. Casino employees boarded the bus, taking the cards from the seniors and giving out the buffet certificates.

Haywood and Lilly entered the casino, following the seniors on the escalator downstairs to the lunch buffet. They enjoyed the free lunch and returned to the escalator to enter the casino. Haywood couldn't believe how many slot machines were lit up and making noise. They continued on the next escalator going to the second floor in the casino. He held Lilly's hand as Lilly told him, "We're only spending fifty dollars on the slot machines."

He laughed and replied, "Yes, let the games begin, as I can't wait to get back home and get the other fifty dollars." Lilly smiled, pointing at the penny machines. She was soon playing them, while Haywood decided to play the nickel slot machine. He kept playing the machine, waiting for the three pears to show on the screen. That's what he needed to win the $100,000 jackpot. All the other symbols showed on the screen except for the three pears, and he quickly lost twenty dollars. Enough with that machine, he thought, moving to the quarter machine. The lit-up ad reporting a win of one million dollars on this one. He placed another twenty dollars into the machine and quickly lost fifteen of it. He spent several hours on the machine after winning small amounts and a Canadian fifty dollar bill. This made him happy, as he now had over fifty-five dollars and would win the fifty dollars at home.

Lilly was still playing the same penny machine.

"Hey, sweetheart, guess what?" He said. "I was playing the quarter machine and got a Canadian fifty, and I still have another five dollars."

Lilly shook her head, opening her fanny pack to show Haywood several Canadian fifty-dollar bills.

Haywood laughed. "We still have to change the currency to American and board the bus, so we better head out."

They boarded the bus at 4:30 p.m. and were surprised to see the seniors on the bus already. At 4:45 p.m., the bus driver walked onto the bus, counting all the seniors including Lilly and Haywood. The bus left the casino, moving toward the United States tunnel. Arriving at the customs gate, which was closed, the bus driver opened the bus door. A United States custom officer boarded the bus, checking seniors' passports, driver's licenses, and birth certificates. Everyone's documents were being carefully checked. The customs officer approached Haywood and recognized him while looking at his passport. This customs officer had worked with Haywood on a criminal case of robbery and kidnapping as the criminals attempted to cross into Canada from the United States. Haywood smiled, extending his hand to the customs officer.

"It's been a long time since I've seen you, Haywood," the customs officer said.

Haywood stood and shook the officer's hand. "Mac! What a nice surprise," he said. "I finally retired from the FBI. I'm enjoying every minute of my retirement." He gave Lilly a hug and a kiss. "This is the real reason I retired and why I am enjoying it. This is my wife, Lilly."

The customs officer shook her hand, not requesting any identification. "Oh yes, I can see why you are enjoying your retirement."

"Be careful and stay safe," Haywood said.

Mac continued checking the rest of the seniors on the bus and waved at Haywood and Lilly as he left the bus.

The bus moved through Detroit on the interstate. An older man sitting behind Haywood tapped him on the shoulder, and Haywood turned in his seat, seeing the man, who was partially bald with gray hair. The old man's wife was sitting next to him, wearing a hair wig and lots of makeup. She was clutching a large black purse. The older man spoke. "Were you a customs officer? I heard the border customs officer talking to you, and he seemed to know you. I thought you might have worked as a customs officer."

Haywood was trying to decide how much he should tell him. "I recently retired from the FBI, and I've known Mac, the customs officer, since we worked on a criminal case together some time ago."

The old man looked at his wife. "See, I told you." He pointed at Haywood. "He was a customs officer."

The wife glared at the old man. "You don't listen, Jim. That man you were pointing at said he retired from the FBI." The old man returned the glare at his wife.

Haywood nudged Lilly. "Sweetheart, when we get older, I hope we don't act like the older couple behind us." Lilly turned in her seat, looking at the older couple and agreeing with Haywood. She kissed her tall, good-looking husband.

The bus continued moving along the interstate in the Detroit city limits. Lilly pointed to some old homes from the bus window, whispering to Haywood, "Honey, how can people live in these poor areas?"

Haywood replied, "Sweetheart, they are just people like us, living like us." "No, they're not like us," Lilly said.

Haywood looked at his wife and whispered, "You just gave me a great idea. Do you want to hear it?"

Lilly began laughing and replied, "You got us here to the casino with your last idea. I won extra cash plus another fifty dollars from our budget, so I would love to hear your new idea."

Haywood hesitated a minute, then spoke. "I'm going to write a book about people who live in those areas you're concerned about. My book will prove to you that they are just like us."

He waited for Lilly's reply, but she only looked at him and smiled. It started raining, and Haywood wondered if gasoline would be found in another parking lot.

RED, the homeless man, was catching the rainwater flowing from the building roof into the barrel. The sky was dark, it was windy, and the tree branches were hitting the building roof. Red had found some candles in Waxxer's Candle Store's dumpster. He lit one candle and walked to a room in the rear of the old, dry building, entering a huge metal structure and closed the heavy door.

CHECKING THE NEIGHBORHOOD

T HE NEXT DAY, HAYWOOD decided to take a trip back to Detroit. He didn't want to wake Lilly. He started the car, and the gas meter showed the tank was almost empty. He drove to the closest gas station and filled up his gas tank. He couldn't believe the price. He put in seventeen gallons at $3.50 per gallon, which totaled $59.50. He missed Lilly not being with him, but knew she would be upset with the new gasoline prices.

While driving to Detroit, Haywood thought about how he had met Lilly. He was a new FBI agent in Detroit and had stopped to get a cup of coffee. It was raining as he raced inside, almost knocking down the woman standing by the coffee shop door and causing her to drop the plastic coffee cup she had been holding. Haywood apologized several times, seeing her spilled coffee, and offered to get her a new cup. Lilly smiled and agreed. They had both sat at a table in the coffee shop, and Haywood ordered two cups of coffee with two beef sandwiches. This woman was so beautiful, and he wondered if she was meant to be his. After ordering the food, he introduced himself. "I am so sorry. My name is Haywood Runyan."

"I'm Lilly Commer, and I am sorry I stood in the doorway because of the rain and no umbrella. But I'm also glad because it's been great being with you. I promise I won't stand in anymore doorways. Why you were in such a hurry? Was it the rain?"

Haywood couldn't believe this was happening as he looked at his watch and answered, "I'm a federal FBI agent," showing her his badge, "and I have a court case in about forty-five minutes."

Lilly enjoyed her sandwich, asking, "How long have you been in the FBI?" "I just finished my training and going to court on my first case of robbery at a city bank," he said.

"Can I go to the federal court with you?" she asked taking a sip of coffee. "It's Friday, and I got out of college early today." Haywood remembered how Lilly stood up from the table, explaining she was a university college professor and would love to hear a criminal case. Haywood was still thinking about Lilly and how they met. The federal judge in the criminal court had set trial for another date so the defendant could have an attorney present. This gave more time for Haywood and Lilly to be together.

Haywood pulled up to Lilly's apartment driveway. She opened the car door, about to step out of the car, when Haywood took her hand and said, "Lilly, it's been great being with you, and I would love to be with you again."

Lilly smiled, replying, "Haywood, thank you for the coffee and the sandwich and for allowing me to go with you to a criminal trial. I would love to see you again. Here's my phone number."

Soon after their first meeting, they were married. Haywood couldn't believe how fast forty years had passed after their wedding.

ARRIVING in Detroit, Haywood stopped in front of the St. Reba Catholic Church. A priest was walking down the church steps and approached him.

"I am Father Pario. What can I do for you, sir?"

Haywood smiled and shook the priest's hand. "My wife and I were on a bus yesterday, traveling through the Detroit area, and she was looking at this neighborhood. She told me 'I don't know how people can live here.' I explained to her that people who live here are the same as us, and she disagreed."

Father Pario laughed. "Sir, you should have listened to your wife."

Haywood took a step back. "I told my wife that she had just given me an idea for a book and that's why I'm here. Do you know anyone I could write a story about who lives in this neighborhood?"

Father Pario pointed across the street at an old standing gate surrounded by high grass. "A homeless old man lives somewhere in the woods. I talked with three of our church kids who threw water balloons at him yesterday as he was running away. I took the kids to the path leading into the woods to apologize to the old man, but he never showed himself. That's all can tell you. He is old and homeless, and he comes out of the woods often."

Haywood thanked Father Pario and returned to his car. Looking at his watch, he decided to wait and hoped to see the old man.Barney Nelson, an employee at Fred Roll's Restaurant, was carting bags of trash to the dumpster behind the building. He was about to pitch the trash into the dumpster when he heard a noise and saw an old man standing there. Barney placed the bags next to the dumpster and ran into the building, hollering for Fred the owner. Fred nodded and laughed, throwing off his apron, and raced to the dumpster. He arrived as Red was climbing over the side and picked up the bags Barney had placed next to the dumpster. Red started walking away, and Fred grabbed him and took the bags. Red looked down. He wouldn't look at Fred.

"How many times have I found you in this dirty dumpster getting food to eat?" Fred asked.

Red shuffled his feet, acting nervous.

"You've been in this dumpster several years, and we see you on our cameras," he told Red. "But every time I get out here to catch you, you're gone. Not today. Let's see what you have in those bags." Fred peered into the bags and raised his voice. "Red, you can't eat this food! It's garbage!"

Fred Roll was a short, thin man, almost bald, wearing glasses. His assistant manager, Sally Nelson was Barney's mom. All the employees enjoyed working for Fred Roll.

The restaurant was a one-story stone and brick structure. Windows surrounded the front and side of the building. Some windows were advertising special meals and desserts. Tall trees offered shade in the parking lot.

Barney Nelson returned to the parking lot, telling Fred, "That's him, he's the old man I saw inside our dumpster."

Fred shook his head, almost losing his glasses. He replaced his glasses and looked at Barney. "This is Red."

Barney looked at Red, holding his nose and hollering, "Hey, old man, you stink. Look at your clothes! When's the last time you had a bath?"

Red was shaking, and Fred raised his voice. "Barney, you get back in my restaurant and get to work now! I will handle this problem."

"Okay, Mr. Roll, I'm sorry," Barney said.

Fred looked at Red. "Barney's a young kid and didn't mean what he says. You stay here. These bags are going in the dumpster. I'll bring you some good food from my restaurant, but don't expect this every time you come back here," he said.

Red found a shady spot under a huge tree to the side of the parking lot while waiting for Fred, who returned with a clean paper bag.

"Here's some salad, a couple meat sandwiches, and a cup of ice cold lemonade for you, my friend," Fred said.

Red stood up, thanking Fred. He then reached into his torn pants pocket and handed Fred some green pellets. Fred looked a little confused by the offering, but smiled and walked towards his restaurant, throwing the pellets into the dumpster as it started raining.

Red sat, enjoying his food while being protected from the rain by the large trees. When he finished eating, he stuffed the plastic container inside the paper bag and left it by the tree.

Haywood looked at his car windows as the rain poured down on them. He was glad he was in his car. Looking at his watch and seeing the time was now 1:00 p.m., he realized he had been waiting for the old man to appear for several hours. He was hungry, so he decided to find a place for lunch. He pulled into the parking lot at Fred Roll's Restaurant and went into the building. While standing there waiting for someone to help him, he heard Fred and Barney talking about Red.

"I happened to overhear you talking about the old homeless man in your dumpster," he said to them.

"Are you a family member of Red?" Fred asked.

"No, I was going to write a story about him after talking with the priest at the St. Reba Church down the street," Haywood said.

"Are you an author?" Fred asked. "Not yet," Haywood said.

"That old homeless man is sitting under a tree in our parking lot, eating the food our boss, Fred, gave him," Barney said.

Fred told Haywood to come with him to talk to Red. As they walked outside, they smelled gasoline coming from the dumpster, and Fred couldn't believe what he saw. His dumpster was filled with gasoline and it was overflowing into the parking lot.

"What the...?" Fred pointed toward the tree where Red had been sitting. "He was over there. Excuse me. I have to do something about this gasoline." Haywood returned with him to the restaurant. Fred called the Fire Department.

Soon the fire department and the city police were at the restaurant. The fire department put foam on the gasoline in the parking lot and the dumpster. The police were keeping curious customers and people in the area from trying to see what was happening when Haywood removed the sack with the plastic container and the cup to his car. He drove out of the parking lot, heading for the FBI office where he had recently retired.

Barney was jumping up and down in excitement. "I can't believe someone would dump and waste expensive gasoline by pouring it into our dumpster."

Fred nodded his head while watching the firemen at work. The police were still keeping the people from the area. The media was now there and reporting about the gasoline. Several people stood in the area with empty gasoline containers.

A state hazmat agent, Jake Broady, asked to talk with Fred Roll, the restaurant owner. He showed his badge and credentials to Fred and asked, "How did the gasoline get into the dumpster?"

"I have no idea," Fred said.

"I took garbage bags to the dumpster and found an old man in it," Barney contributed.

"I almost forgot about that," Fred said. "We have cameras around our parking lot and the dumpster."

Agent Broady nodded. "Do you have the type of cameras where pictures are on a recorder?"

"No, but our cashier may have seen someone while looking at our monitor screen," Fred said. "Let's talk to her."

Sally Whylick was Fred's cashier. She was a pretty woman in her late fifties, tall with long brown hair. She had gold-rimmed glasses and looked very professional. Fred introduced her to Agent Broady, who asked, "Have you been watching the camera monitor?"

Sally nodded her head, swishing her long brown hair behind the right side of her ear. "I've seen four people by our dumpster. Barney Nelson took some bags of garbage out, and I also saw an old man in it, then our owner, Fred, with someone else, and that's all I've seen today."

"Whose this old man?" Agent Broady asked.

"That's Red. He stays somewhere close to here," Barney said.

Fred raised his voice. "Red didn't have any gasoline with him. He comes around here often, getting into our dumpster looking for food."

"Someone around here must know about this, and I will find him," Broady said. "Who was the other person with you?"

"I don't know his name or where he's from, but he wanted to write a book about Red after talking to the priest at the St. Reba Church," Fred said.

Agent Brady thanked everyone and left in his van.

Haywood looked at his watch and called Lilly on his cell phone while traveling to the FBI office. "Hi, sweetheart. I am here in Detroit, investigating that story I told you I was going to write. I talked to a Catholic priest, Father Pario, who told me I should have listened to you."

Lilly laughed. "I agree. You should have listened to me. So when are you coming home?"

"I've found an old man, homeless, living in the woods, and in the neighborhood you were concerned about. I have his fingerprints."

"Why? What did he do?"

"I don't know if he did anything. I just want to know who he is so that I can find him and talk to him," Haywood said. "I'm heading to my old office now to see if I can run his fingerprints." He hesitated.

"And please don't fix dinner. When I get home, we can eat out. I may be a little late tonight. Love you."

"I love you. Please be careful," Lilly said.

AGENT Broady called Mike Trident, owner of Trident's Grocery, from his state hazmat office.

"Mr. Trident, several days ago you had gasoline in your dumpster and parking lot behind your store. Was there an old man in your dumpster?"

Mike Trident thought a minute and answered, "Yeah, you must be talking about Red."

"Why was Red in your dumpster? Did he have any gasoline with him?" Mike laughed. "You got to be kidding me! That old man could never carry that much gasoline! I need to get back to work. Red doesn't know about the gasoline puddles in our parking lot or gasoline in our dumpster."

"I'm coming to your store to talk with you about this old man you call Red," Agent Broady said.

Mike pounded his desk with his fist and cursed. "Hey man, I have two new bills I have to pay because of the mess from that gasoline. You still haven't found the culprit, have you?"

"We're working on it," Broady said. "In the meantime, if you find Red, please call me any time during the day or night. I really need to talk to him for information. We will catch the culprits, I promise, but I need your help."

Mike laughed. "All right Broady, whatever you say. We believe Red lives in the woods somewhere around here."

HAYWOOD carried Red's paper sack into the FBI office. He was thinking about how nice it would be to visit with his former boss, Ted Ruffen, and his staff. He walked by his old office and wondered if Ted was still there in Detroit. He opened Ted's office door, seeing the surprised look on Ted's face. Haywood laughed and walked toward Agent Ruffen. Ted jumped up from his oak desk, extending his hand to Haywood.

"I told your secretary I wanted to surprise you," Haywood said, shaking his hand.

"How's retirement, you old fart?" Ted said, laughing. He stood several inches taller than Haywood—a thin man with a short beard and mustache. He put his hands up above his head and asked, "Do you still think you can put handcuffs on me?"

Laughing, Haywood gently pushed Ted away.

Ted offered Haywood a chair and returned to his desk. "What have you been doing since your retirement? Have you been working on any good federal cases?" He looked at the paper bag in Haywood's hand. "Did you bring me something to eat?"

Haywood told Ted about the trip with Lilly on the bus to the casino in Canada and how, on the return trip through Detroit, Lilly had given him the idea to write a book about the Detroit neighborhood. He told him that he'd talked with a priest at St. Reba Catholic Church, and their discussion about the old man who lived in the woods across from the church. Haywood then mentioned the gasoline puddles in Fred Roll's Restaurant parking lot and dumpster.

Ted left his desk and closed his office door. Then he turned on the office intercom to his secretary. "I am busy and do not want to be disturbed," he told her. Then he said in a low voice to Haywood, "We're concerned about this gasoline found in parking lots and dumpsters in Detroit. So far it has only happened in this city. We are now believing that a local terrorist group may be doing this to check our response time. We have a meeting planned for next week in our conference room with agents from the state hazmat, ATF, government homeland security, the economy commissioner, the city fire chief, and the city police chief."

"How about checking the plastic dishes and cup in this sack for fingerprints?" Haywood said.

Ted set the sack on his desk. "Why? Does this have something to do with the gas?"

"It might," Haywood said. "Check it out and let me know what you find out, okay? You owe me one."

Ted nodded. "You old son-of-a-gun, you're up to your old tricks again.

Aren't you supposed to be retired?" Haywood shrugged sheepishly.

Ted laughed. "Okay. I'll run the prints and let you know." "Thanks."

"How's that pretty wife of yours handling your retirement?" Ted asked.

"Thanks for asking about Lilly," Haywood said. "She's doing great. I hope when you retire you'll enjoy it every day like me."

He shook Ted's hand again before leaving.

Haywood was glad he had called Lilly earlier because he planned to discuss his trip to Detroit with her while they dined out.

"It's good to be back home, sweetheart," he said, giving her a hug and kiss. "Where do you want to eat? My treat."

Lilly smiled. "I was going to use the money I won at the casino." "Nope," Haywood said, embracing her again. "My treat this time." They went to a local buffet restaurant and began filling their plates with food. Haywood was glad that Lilly liked the buffet because of the variety of food and drinks all at one price. They sat at a table eating while Haywood told Lilly about his trip to Detroit. He returned to the buffet several times to fill his plate.

"Hungry?" Lilly said, smiling.

"Sweetheart, I haven't eaten all day. Don't worry. I won't let any of this go to waste."

INVESTIGATION OF RED, THE OLD MAN

S TATE HAZMAT AGENT JAKE Broady arrived at Mike Trident's grocery store and showed his credentials to Mike. The agent then started asking questions.

"Wait," Mike said, stopping him. "I've got to get these bills paid. First, my employees keep coming in and interrupting me, then you show up asking all these questions. Please, sit here and let me finish what I was doing. Then I'll answer your questions."

Agent Broady sat and watched Mike write out checks and place the checks and bills in addressed envelopes. Mike finished with his paperwork and thanked Agent Broady for his patience.

"Mike, tell me about the old man who was in your dumpster before finding the gasoline in your parking lot," Broady said.

Mike tapped a pen on his desk. "Red's been in our dumpster looking for food for who knows how long—maybe years. All I know is that he's old, probably in his late seventies, wears dirty clothes, stinks, and has stringy red hair. He's also very shy—maybe mentally slow." Mike paused, scratching his head. "We call him Red because of his red beard and mustache. But no one knows where he stays. He just seems to appear and then disappear. You're wasting your time looking for him."

Agent Broady stood up. "Thanks, Mike. If you see Red again, please keep him here and call me."

He handed Mike his business card and left.

Haywood was watching the weather station report on a severe storm when their home phone rang.

"Your former boss is on the phone," Lilly said, handing Haywood the receiver.

Haywood heard Ted Ruffen's voice on the other end, "Haywood, are you sitting down?" Lilly saw Haywood's puzzled look and went into the other room, picking up the phone to listen.

"We got identification from the fingerprints on the plastic dishes you gave me," Ted said. "Are you sitting down?"

Haywood was still standing, but he said, "Yes I'm sitting. What'd you find out?"

"The fingerprints belong to a person named Oley Washington Jr. We ran the name through our database in Washington, DC and got a hit in the United States Air Force. Oley Washington Jr. was a captain in the Air Force. He was considered one of our best fighter pilots. His plane was shot down somewhere near the North Vietnam border and he was captured and thrown in a Vietnam prison war camp."

Haywood grabbed a pen and paper, scribbling down notes while Ted talked.

"Oley Washington Jr. helped many of our soldiers escape from the prison camp, but he disappeared after coming back to the United States. Our military has been looking for him for many years. The president of the United States was going to present him with the Congressional Medal of Honor. Records show he was staying at a hotel in Washington, DC with his parents a few days before he was to receive the Congressional Medal of Honor and the Washington City Police have a report about him being taken to a nearby hospital with head injuries from a beating. No one was ever arrested in connection with the incident. Oley left the hospital and then disappeared. It was reported he had amnesia. He has never been heard from since."

Lilly, who was still listening on the other phone, let out a short cry before catching herself.

"What was that?" Ted asked.

"I have to go," Haywood said. "Something's wrong with Lilly. Thanks for the information." Haywood hung up the phone and looked over at Lilly who had come back into the room. She was sitting on the couch crying. Haywood tried to tell Lilly about the phone call, but Lilly just continued shaking her head and crying.

"Sweetheart, what's wrong?" Haywood asked. "Are you hurt?" Lilly shook her head. "I need to lay down." "Lilly, sweetheart, please talk to me about why you are so upset. What is going on?"

Lilly didn't answer him, but lay down on the couch. Haywood went to their kitchen, returning with a glass of ice water. She sat up and drank the water, still crying.

"Sweetheart, please tell me why you are so upset."

Calmer now, Lilly said, "Honey, I'm so sorry. I don't want to talk about this now, but maybe later."

Haywood was shocked. What could have upset her so much?

AGENT Ted Ruffen was in the FBI office at his desk when his secretary announced that US Air Force General Mike Golden was on the phone.

"Agent Ruffen, have you located Oley Washington Jr. yet?" Ted replied, "We're still working on that problem, general."

"I have more information for you about Oley," the general said. "His father was Oley Washington Sr. He was a great scientist working for our military on new forms of energy. Agent Ruffen, do you know about a top secret project called Blue Book?"

"Are you referring to a study on space creatures in flying saucers?" Ted asked.

"Yes," the general said. "Oley Sr. was doing great work until his son disappeared. Several months later Oley Sr. committed suicide due to being depressed about his son's disappearance. Oley Sr.'s wife is still living in Florida." General Golden paused, then continued. "We talked with Oley's mother, who couldn't believe her son had been found. She believed her husband and son were together in heaven. Oley's mother begged us to bring Oley Jr. back to her, so we need to find Oley. He was never dismissed with honor from the Air Force. We need to prove Oley left the hospital with amnesia or left the military with an unauthorized leave. So please call me when you have him in custody."

"We'll find him," Ruffen said. "I'll call you when he's here in our office."

Ted hung up the phone, and his door opened. His secretary stood in the doorway smiling at him. He looked at her, shook his head, and picked up the phone to call Haywood Runyan.Lilly Runyan was still lying on the couch when the phone rang. Haywood answered it. Ted Ruffen's voice was excited on the other end.

"I just talked to a General Golden who gave me more information concerning Oley Washington Jr. You may want to sit down again."

Haywood smiled. "I'm sitting down." Lilly had stopped crying and was now sitting on the couch wondering why her husband was standing when he said otherwise.

Ted told Haywood about his call with General Mike Golden. After hanging up, Haywood walked over to Lilly, shaking his head. "Ted wants me to find Oley Washington Jr. and bring him to the FBI office in Detroit. Since you're sitting up, will you tell me why you've been so upset? You know I love you; you can tell me anything."

Haywood sat next to Lilly and held her. "Sweetheart, what are you scared to tell me? I love you so much. You can talk to me about anything."

After wiping her eyes with a tissue, Lilly said, "Honey, I knew Captain Oley Washington Jr."

Haywood looked at her, puzzled. "How did you know Oley Washington Jr.?"

She continued, her voice trembling. "God please forgive me. Honey, I was there with the college war protesters. Our group saw Oley Washington Jr. in his military uniform carrying a bag and walking from the hotel. We approached the captain, surrounded him. I tried to stop them from beating him, but they wouldn't listen to me." Lilly paused to blow her nose, then continued. "He fell to the ground, and several protesters kicked him and hit him in the head with their fists. I wish I knew who they were because I would have reported them to the police. Honey, he was dragged in an alley and left alone. The group ran from the area. I tried to help him, but I thought he was hurt bad, so I called an ambulance. But when I saw the ambulance coming, I left the area." Lilly looked at Haywood. "That poor man. I am responsible for the captain's condition!"

Haywood attempted to calm Lilly, but she continued. "I was bad and foolish. Oh, God, please forgive me."

Haywood held Lilly tight. "Sweetheart, it's going to be okay. That man could have been someone else. You don't know for sure it was Oley." Looking at Haywood, Lilly shook her head. "It was him. I saw his name on his uniform. Oh, God, please forgive me. Honey, please forgive me."

Haywood held Lilly tight, kissing her, hugging her, and telling her everything would be okay.

FATHER PARIO'S BAPTISM
AT ST. TREBA CHURCH

BEFORE COMMUNION ON SUNDAY, Father Pario approached the altar and looked up at the high ceiling, praying a blessing over the congregation. Then he kissed the large crucifix that was handed to him. He walked to the baptismal font beside the altar and requested the new catechumens who were ready to be baptized to rise and come forward. Seven adults joined him at the font. The baptismal was the size of a small pool—much larger than a normal Catholic baptismal font. It was a relic from the nineteenth century, made of pure marble with an angel sculpture above that poured water into the pool.

"What do you ask of God's Church?" Father Pario asked the catechumens. "Baptism," they responded in unison.

He made the sign of the cross on each of their foreheads, read a passage from the Bible, then rubbed a bit of anointing oil on each of their necks. Then he said a blessing over the water.

"Do you renounce Satan? And all his works? And all his empty promises?" he asked them.

"We do," the catechumens replied.

"Do you believe in God, the Father Almighty, Creator of heaven and earth?"

"We do."

"Do you believe in Jesus Christ. . . ? Do you believe in the Holy Spirit. . . ? "We do."

With the vows completed, Father Pario motioned for the catechumens to line up. The first in line—a middle-aged woman—approached and held her head over the basin as Father Pario poured a bit of water over her head. After each took their turn, the priest lit a candle for each new convert. They stood holding the candles, as he said a blessing over them and their families. Once the baptisms were over, Father Pario said the Mass and then carried in each hand the chalice of wine and the chalice of the host. Before he reached the line to give out communion, however, a deacon approached him.

"Father," the deacon whispered, "there's a strong gasoline smell coming from the baptismal. Look, the water coming out of the angel is gold in color."

Father Pario marveled at the site. Several church members who were sitting in pews close to the baptismal left their seats and headed for the exits. Father Pario stopped the service, asking the congregation to leave the building, then he went to his office and called the city fire department. As the congregation members stood in the church parking lot watching the fire department and police arrive, Father Pario directed the fire and police into the church. The media arrived, and the story was soon being broadcast on television and radio. People began showing up from all over with empty gasoline cans.

Father Pario opened church windows with the help of Father Steven and hollered to those in the parking lot. "All is okay. Everyone can go home. The fire department and police will handle our problem. God bless you."

No one left.

The city fire chief walked over to Father Pario. "The gasoline appears to be very clean. You could probably use it in your car. But we can put foam in your baptismal to soak it up if you want."

Thinking about how high gasoline prices were, Father Pario decided to save the gasoline for St. Treba church members. *This must be a gift from God*, he thought.

He walked outside in the church parking lot and made his

announcement. "God has given our church members gasoline for their cars. I have talked with the fire chief who has agreed to let our church members remove the gasoline. You will need the correct gasoline containers before removing the gasoline. Several fireman will check your containers before you can remove the gasoline. I've noticed many non-church members here in our parking lot with gasoline containers. I am sure you have heard about the gasoline in our church as reported by the media. I'm sorry but all non-church members will need to wait until our church members get the gasoline first. This is God's gift to our church. Does anyone have any questions?"

All was quiet, and Father Pario returned to the baptismal so that he could verify that all those lining up for the gasoline were church members.

State Hazmat Agent Jake Broady, ATF Agent Henry Tiffer, FBI Agent Ted Ruffen, and c showed their credentials as they arrived at the church. They stood beside the baptismal, smelling the gasoline and watching church members filling their containers. Father Pario smiled. "This shows God's love for our church members and to our church. God has supplied gasoline for our church members and other people."

"Father Pario, was an old homeless man called Red getting baptized?"Agent Broady asked.

Father Pario shook his head. "Red isn't a church member and has never been in this church. He lives across the street beyond the gate, somewhere in the woods."

Ruffen told them about how retired Agent Haywood Runyan brought the plastic dishes to his office for fingerprints.

"The fingerprints belong to Oley Washington Jr., a hero in the Vietnam War and a captain in the Air Force." Ted explained the story about Oley's mysterious disappearance and how his mother was hoping to find him.

"Guys, someone is placing gasoline in our water, and we have to find them before they hurt or kill our American people," interrupted Agent Broady.

Kyle Hampter obtained a gasoline sample to study down at headquarters. As the baptismal was being emptied, it seemed to be replenishing itself, and people were filling their gasoline containers and racing to their cars to fill their fuel tanks then return to the baptismal for more gasoline. Finally, after what seemed like hours, the flow of gasoline stopped and the church baptismal was empty. Everyone, including the media, left the church except for some church members who volunteered to help Fathers Pario and Steven scrub the baptismal.

They all cheered when the angel began spraying water again. "What a miracle!" Father Pario chanted.

Father Pario asked several church members to bring fans to blow the gasoline smell out of the sanctuary.

"Well," said Broady to the other agents as they were leaving, "maybe Kyle will have some information for us about the samples he took." Lester King, owner of East and West Coast Global Oil Industries, was talking on a conference call with his company representatives. He was concerned about a hurricane reported to arrive on the East Coast and the effect it would have on the price of gasoline. "Gentlemen, if this hurricane destroys our refineries, we will have great shortages of gasoline for our consumers. We should raise our gasoline prices now." A long pause continued ensued.

"Gentlemen, I am waiting for your answers," King said, placing his legs on top of his desk and leaning back in his chair.

Representative Kyle Hampter, who was invited on the call, spoke up. "Sir, I may have a solution," he said.

Kyle told Lester King about the gasoline found in the St. Treba baptismal and the samples he took. "We tested the samples. The gasoline is the best grade known to humankind. As crazy as it sounds, we believe the water produced the gasoline—"

"Are you telling me someone has discovered how to turn water into gasoline?" King interrupted.

There was a long pause. Lester King continued. "Do you reps realize this information could ruin our business? You must find out who is responsible for this immediately! I don't care what the cost or what you need to do, but you must find them and stop them. I will be waiting to hear from you about this problem. Before I dismiss this conference call, I would like for Kyle Hampter to call me. That's all for now, so let's all get back to working and raising the gasoline prices before we have this hurricane, which will be our excuse. Agent Hampter, please call me now, and my thanks to all for attending this conference call."

Lester King had just lit a cigar when his secretary entered his office, telling him that Kyle Hampter was on the line. Lester picked up the phone.

"Mr. Hampter I need your help and I can pay you whatever you need to find out who is responsible for turning water into gasoline. It's unbelievable. But if it's true, do you understand what this can do to my industry? The Saudi oil cartel and other foreign oil companies will be affected as well. Please find the person involved as you have the contacts. I need your help to get rid if this problem now. I will pay your price."

After a long pause, Hampter replied, "I'll see what I can do. I'll be in touch.

HAYWOOD RUNYAN AND
OLEY WASHINGTON JR.

AFTER TALKING WITH FBI Agent Ruffen, Haywood decided he needed to return to the neighborhood in Detroit to find Oley Washington Jr. Haywood thought Oley needed to know about his mother, who was living in Florida, and his father, who was deceased.

He decided the best place to start was at St. Treba's. He found Father Pario talking with Father Steven near the confessionals. Haywood greeted them and shook their hands.

"I heard about your episode in the church yesterday. It was all over the news," he said.

Father Pario nodded. "Ted Ruffen was here and told us about Oley Washington Jr. I had seen him the other day leaving the woods and approached him, but he never looked up. This was before I knew he was Oley. I called him Red and apologized to him for the water balloon incident. I almost completely forgot it, but he gave me some small pellets and told me to put them in water and wait."

Haywood perked up. "What kind of pellets? What'd they look like?" "Kind of like green jellybeans," Pario said. "They were about that size." "Father, where are those pellets now?" Father Steven asked.

Father Pario gave them both a smile. "I remember now. I put them into my robe pocket. It's hanging in the closet." They followed him into the priest changing room behind the sanctuary. He removed his robe from a hanger and checked the pocket. No pellets.

"This was the robe I was wearing during the baptisms. They must have fallen out of my pocket and dropped into the baptismal while I was baptizing our catechumens."

He paused again in thought. "Oley told me to put them in water and wait. They were weird greenish-yellow looking beans. Those pellets must have caused the water in our baptismal to turn into gasoline. And here I thought it was a miracle from God."

They left the church office and walked to the baptismal. As they watched the angel pouring water into the pool, Father Pario whispered, "No one must ever know about the pellets changing this water to gasoline."

"No one must ever know what we know now about Oley," Haywood whispered back.

WHEN Haywood returned to his car he noticed Oley walking toward the gate leading into the woods. He followed Oley, being very careful to leave the gate open to avoid any noise. Oley climbed over brush and entered a concealed old brick structure. Haywood crept over the brush like a hungry lion sneaking up on its prey. He slowly walked toward the building entrance and couldn't believe what he saw. The building's roof was covered by tree limbs and was torn open in different areas. It was starting to rain, and water from the roof was pouring into an old barrel. Haywood stood in a room covered with trash. He slowly walked through another room with several holes in the floor, peering into the next room through a broken doorway. Oley was there, bending over and lighting a candle. Slowly he walked into the room.

Shocked to see Haywood, Oley knocked over the candle and quickly looked away. "Please," he said, shaking. "I just want to be left alone."

Haywood approached Oley and slowly crouched down and extended his hand. Here was a real hero, a captain in the Air Force who was supposed to have received the Congressional Medal of Honor from the president of the United States of America, but Haywood now was looking at a man covered in filthy rags, shaking in fear like a cornered animal. A long, dirty beard covered most of his face and his long hair was so dirty you could hardly tell it was red. Haywood reflected that he had never seen a man this ragged and filthy before. He was so thin he appeared to be starving.

"I just want to be left alone," Oley said again, ignoring Haywood's extended hand.

Haywood remembered about Oley's amnesia. "Oley Washington Jr. Is that your name?" He spoke softly. Oley moved away from Haywood. "Please leave me alone."

Haywood watched Oley leave the room and go into a darker room with a huge steel structure in the shape of a cube. Oley went into the cube and closed the door with a loud bang. Haywood picked up the lit candle that Oley had left and followed him. He approached the metal door, realizing it was an old bank safe. He grabbed the handle, turning it, and almost dropped the candle when the door opened. Oley stood there facing him and Haywood extended his hand again. Instead of shaking it, Oley placed several pellets in Haywood's palm. From the candlelight Haywood could see several blankets stacked against the interior wall of the safe. Then he heard a strange popping, humming sound coming from the safe. Haywood looked at the small greenish-yellow pellets Oley gave him and placed them in his pants pocket.

"Put them in water and wait," Oley said.

Haywood nodded. "Are you Oley Washington Jr.? Is that your name?" "I just want to be left alone," Oley said, looking down at the safe floor.

Haywood gave Oley a friendly smile. "Were you a captain in the Air Force?"

That question seemed to startle Oley. He looked up and directed his gaze at Haywood.

"Oley is your first name, correct?" Haywood said, extending his hand. Oley seemed dazed.

"Oley, will you come with me?" Haywood asked.

Oley began backing into the safe when Haywood threw the lit candle into a puddle of water outside of the safe. He grabbed Oley, placed him in an arm lock and pushed him out of the safe. Oley was too weak to fight back.

"I just want to be left alone," he whined.

Haywood had brought along some handcuffs just in case he needed them, and now placed them on Oley. Then he led Oley from the building, over the brush, through the woods and to the open gate. Then he pulled Oley inside the St. Treba Church and into Father Pario's office.

Father Pario was startled at their sudden appearance. "What's this?" he asked, looking at Haywood.

"I just want to be left alone," Oley told the priest. "Father Pario, I apologize for the intrusion and for the handcuffs. But, trust me, this is the safest way I could get him to come with me. This is Oley Washington Jr. He suffers from amnesia. Can you help me get him cleaned up? He needs bath, then a haircut, a shave, and some clean clothes. I will pay you for all of this."

Father Pario nodded. "Not necessary, Haywood, we should have some clothes here."

He removed a small bell from off a shelf and rang it. Father Steven came in about a minute later.

"Please go and find Sister Martha and bring her to my office," Father Pario said.

Haywood continued holding Oley as Father Steven left the office. "I just want to be left alone," Oley said, his whole body shaking.

Father Pario gently placed his hand on Oley's shoulder. "We're trying to help you. We are going to get you cleaned up with some clothes, and you will feel better."

His words seemed to calm Oley, and Father Pario winked at Haywood, who released Oley from the handcuffs. Oley stood with his hands at his side, gazing intently at the floor. "I just want to be left alone," he said again.

Haywood sat Oley down in an office chair and sat down in another chair next to him. Father Steven returned to the office with Sister Martha, who looked at Oley, nodding her head. "Heavens to mercy, what do we have here?"

Sister Martha was sixty-one with a stocky build. She was taller than

Father Steven, with gray hair cut short, dressed in a sweatshirt, pants, and tennis shoes. A gold cross on a golden chain was around her neck. Father Pario ordered them to wash Oley, shave him, cut his hair, and put new clothes on him. Sister Martha took Oley's hand, leaving Father Pario's office with Father Steven walking behind them.

"Father, I have to do something now and will return soon. Please keep Oley here until I return." Haywood said.

He returned to his car and grabbed a flashlight. Then he walked across the street to the gate leading into the woods.

Several kids approached him at the gate. The oldest boy, tall and wearing a baseball cap, seemed to be the leader. "Hey mister," he said, "you better not go into those woods. An old man may cause you serious problems."

Haywood thanked him and returned to his car, waiting for the kids to leave the area. He was thankful they didn't see him remove Oley from the woods to the church. After the kids left the area, Haywood walked around the gate into the woods. He entered the old deserted bank building and went straight to the vault. With flashlight turned on, he opened the door of an inner chamber and marveled at the site. A long string of greenish-yellow pellets were hanging throughout the chamber. They were popping and humming and multiplying before his eyes. Looking around inside the safe, Haywood found an open empty box and put the string of pellets in the box. There was a large military duffel bag resting against the vault wall, so he opened it and found Oley's military clothes, including his Air Force coat with his captain bars attached.

"There's no doubt about his identity now," Haywood said out loud. He returned to his car and placed the box of pellets in the trunk. Then he carried Oley's duffel bag to the church.

Haywood was shocked when he walked into Father Pario's office. Standing next to Father Steven was a new Oley Washington Jr. Although still thin, Oley looked about ten years younger without the beard and long hair. He was wearing new clothes and appeared to be less nervous, except that he was still muttering, "I want to be left alone."

"You did a fantastic job!" Haywood told Father Steven and Sister

Martha. "He looks like a new person. I returned to the deserted bank building where he was living and found this military bag with his Air Force clothes and uniform. His captain bars are still attached to his coat. So, we know for sure that he is Captain Oley Washington Jr. Is it okay if the captain stays here for the time being, Father?"

Father Pario nodded his head. "Let's get the captain something to eat," he said.

Haywood thanked everyone and headed for home.

THE FBI MEETING
AND LESTER KING

S TATE HAZMAT AGENT JAKE Broady was standing in the FBI conference room in Detroit.

"We have a lead on the person responsible for the gasoline problems," he said with excitement. "We've found a homeless man who was present at the Trident Grocery store, Wesswell's Gas Company, and Fred Roll's Restaurant. The man was at each of these locations and left the areas when the gasoline puddles appeared. Agent Ruffen ran his prints and we came up with his identity: Oley Washington Jr., a captain in the US Air Force who was to receive the Congressional Medal of Honor until he disappeared."

Government Energy Commissioner Kyle Hampter shook his head, stood up, and reported on the samples he had taken from the church. "Our lab tested my samples, which proved they were one hundred percent pure grade gasoline. The tests indicated that the gasoline came from water, and the water continues to make more gasoline. Our techs are still trying to understand how this is done. They believe an unknown formula is involved. No matter how it happens, it's important to recognize that this could change our transportation industry. Being independent of foreign oil could help our country's economy. We need to find this formula immediately."

FBI head agent Ted Ruffen stood, thanking Hampter for his information, and adding, "We don't believe Captain Oley Washington Jr. knows how the water is changed into gasoline," Ruffen said. "And we are unsure of how he is connected to all this—if at all. He will soon be here in my office and can speak for himself."

"Agent Ruffen," Broady said, "will you notify me when the captain arrives so that I can talk with him?"

Ruffen nodded.

"I hope you find who is doing this, as these incidents are costing our department time and money," said the Detroit fire chief. The Detroit police chief nodded in agreement and said, "I've invited several guests to this meeting who may be able to help us with our investigation."

He opened the conference room door and in walked Mike Trident of Trident Grocery, Brock Wesswell of Wesswell's Gas Company, Fred Roll of Fred Roll's Restaurant, and Father Steven of St. Treba Catholic Church. The police chief made the introductions before the questions began.

"Father Steven," Kyle Hampter said, "do you have any idea how the water in your baptismal turned into gasoline?"

Father Steven shook his head. "I've just been ordained a priest. I haven't been at St. Treba very long, so I'm not sure if I can be of any help."

"Can you check with Father Pario and let us know if you discover anything new?" Hampter asked.

Father Steven nodded.

Although several agents questioned the new arrivals, no new information was revealed about Captain Oley Washington Jr. Father Steven knew about Oley giving Father Pario the greenish-yellow pellets, but he was told by Father Pario not to say anything at this meeting.

FBI Agent Ted Ruffen stood. "I want everyone to know that we are committed to solving this case and will keep you posted on any new developments. Before we adjourn, I'd like Father Stevens to offer a prayer."

Father Steven stood and prayed over for everyone attending the meeting, including catching the persons responsible for changing the water to gasoline. Everyone shook hands while several Detroit police officers waited to take Mike Trident, Brock Wesswell, Fred Roll, and Father Stevens back to their workplaces.

SEVERAL days later, Lester King met with Kyle Hampter at Borger's Dine and Wine restaurant in Washington, DC.

"It's been several days since we talked," said Lester while sipping a red wine. "What've you found out?"

King reached into his pocket, removed a small envelope, and pitched it on the table toward Hampter. "Go on, Kyle, open it. It's a little reward with much more to come, provided you give me the right kind of information." Kyle picked up the envelope and opened it. Several thousand-dollar bills were inside. His eyes grew large as he counted the money. "Oh boy. I don't believe this," he said.

Raising his voice into a harsh whisper, Lester grabbed the envelope from Kyle. "You earn this by keeping your mouth shut and getting information to me. I want to know how that water turns to gasoline. I want to know who is doing this and who has the formula."

Hampter cleared his throat. "Mr. King, the other day I was in a meeting with the Detroit FBI and other city, state, and federal officials. The FBI agent said the old man who was seen at the locations where the gasoline was found has been identified. Only he's not as old as we originally thought . . . More like in his early sixties. His name is Captain Oley Washington Jr. He was a Vietnam war hero who went missing several years ago after getting amnesia. He seems to be the only connection to this gasoline thing, but they say he's not a suspect."

Lester King cleared his throat and motioned toward two large men sitting at a table nearby. They nodded back. Then King returned the envelope to Kyle.

"That's helpful," King said. "I'll expect more information from you soon. You've seen my men, so don't try to cross me." He gulped down the rest of his wine and added, "Your dinner's on me." Then he placed a one-hundred- dollar bill on the table.

One of his goons, a large man with a braided ponytail down his back and a diamond-star earring dangling from his right ear, grabbed

King's coat from the coat rack. Kyle noticed a bulge protruding from the man's right hip as he helped Lester with his coat. The other man was shorter with a small mustache and a tweed jacket. He stood where Kyle could see the shiny handle of his revolver in his shoulder holster. The two men walked on each side of King as they left the restaurant.

HAYWOOD USES THE PELLETS

AYWOOD'S CAR HAD BEEN sitting in the garage for several days after he had returned from Detroit. Remembering the string of pellets he had left there, he slowly opened the trunk and couldn't believe what he saw. The string had grown into a large glob of pellets that almost filled the entire compartment. Haywood removed one of the pellets from the glob then slammed the truck closed.

Lilly, who heard the commotion, peered into the garage from the door window. She saw Haywood putting water into a gasoline container, but she didn't see him drop the pellet into the water. Then she saw Haywood putting what she thought was water into the gas tank.

"Honey, what are you doing?" She said, bursting through the doorway and almost falling on the steps.

Haywood grabbed her and helped her down the steps and into the garage. He showed her the gasoline container containing only gasoline. Lilly laughed. "I thought I saw you put water into that container. Did you know I was watching and decide to pull a joke on me?"

Haywood held Lilly tight, then whispered, "Sweetheart, I'm going to show you something." He opened the car trunk. The greenish-yellow pellets hummed and popped in the huge glob. Before Lilly could say anything, Haywood dropped a pellet into a can of water, which shortly became gasoline.

Lilly was speechless.

"Remember when I went back to Detroit to find Oley Washington Jr. for Ted Ruffen?" Haywood said. "I didn't tell you about these pellets that I found in the old bank where Oley was living. When you put the pellets in water, they turn the water into gasoline."

"That's amazing," Lilly said. "It's a real miracle!"

"Yes, and so far there are only a few of us who know about it. I need to see Father Pario and my former boss."

Lilly kissed Haywood. "Please take me with you," she said. "I want to see Oley and ask him to forgive me for what happened to him in Washington, DC."

Haywood agreed. He found a large wooden crate and placed the pellets inside. Then grabbed a shovel and dug a hole in their backyard and buried the crate next to a few bushes. When he was satisfied that the area looked undisturbed, he returned to the house.

THE next morning Haywood and Lilly headed to Detroit. Fathers Pario and Steven were finishing breakfast in the rectory when they arrived. After Haywood made the introductions, they headed to Pario's office. Lilly looked over at Father Pario, then walked up to Oley. "Oley, I need to tell you something. I am so sorry for what happened to you in Washington, DC. I was there when you were beaten up by the protesters. I tried to stop them, but no one would listen to me."

Oley was looking at the floor as if not listening, but after Lilly spoke, he suddenly looked at her and then jumped out of his chair. "I remember you! You're the little angel who was trying to stop the group by hollering at them and slapping them. They all ran away, but you stayed with me and helped me. I remember, I remember. Why am I here? Wasn't I supposed to receive a medal from the president? Where are my parents? Where's my uniform?" He gave Lilly a hug and asked, "Where am I? What's going on?"

Father Pario raised his hands in praise to God and Haywood was in shock. Because of seeing Lilly, Oley's memory had been restored. Haywood motioned for Oley to sit down and began telling him what had happened to him and how he was now here in Detroit. Then he introduced him to the priests.

"They've been letting you stay with them since we found out who you were," he explained.

Oley thanked them and said, "My poor parents. I need to call them. Can I use your phone?"

"We'll help you with that," Father Pario assured him. Then he pointed to Lilly. "This little angel is Haywood Runyan's wife. This is Haywood Runyan. Haywood started looking into your story because he wanted to write a book about you."

Oley shook their hands. "Is it okay if I go to my room and change into my Air Force uniform?"

"Certainly," Father Pario said. "Father Steven, will you show him where it is?"

Oley smiled. "I remember the way. I'll be right back."

Haywood gave Lilly a hug. "I don't know what just happened, but I'm pretty sure you brought Oley back his memory. I am so proud of you."

Just then, Sister Martha burst into the room. "I just saw Oley rushing down the hallway to his room. Is something wrong?"

Father Pario smiled and said, "It's a miracle, Sister Martha! Oley knows who he is now thanks to Lilly here."

Lilly was in tears and Sister Martha hugged her. Then Captain Oley Washington Jr. stood in the doorway decked out in his old Air Force uniform. Although the uniform hung loose on him due to his emaciated condition, it still reminded all those in the room that they were in the presence of a real hero. Haywood thought about the pellets and realized Oley might need security protection. Hopefully Ted could arrange that, he thought.

Oley saw that Lilly had been crying and walked over to her, giving her a hug. "Don't cry anymore," he said. "You helped me, and I can't thank you enough."

"Will you please come with us?" Haywood asked. "I'd like for you to meet my former boss, FBI Agent Ted Ruffen. He has some important questions he needs to ask you."

Oley agreed and they walked out into the church parking lot. Lilly insisted on riding in the backseat, so Oley opened the back door for her. Then he sat next to Haywood in the front.

"WHY are we going to the FBI headquarters?" Oley asked.

"That's where Haywood used to work," Lilly explained. "He's retired FBI."

Haywood glanced over at Oley. "I believe you will need security protection until you return to the Air Force."

"Why would I need protection?" Oley asked. "I can't tell you that now, but you'll know as soon as we talk to Agent Ruffen."

TED was surprised to see Oley in his right mind wearing his old uniform when Haywood, Oley, and Lilly walked into his office. He stood up from his desk and shook hands with them.

"This is a nice surprise. It makes my day seeing the Captain in his uniform," Ted said.

"Mr. Ruffen, can you please tell me why I'm here?" Oley said. "Haywood keeps talking about security protection. Why do I need protection?"

"I need to show you something, Ted," said Haywood.

He noticed a pitcher of water on Ted's desk and dropped a pellet into it. The water slowly turned into gasoline. Ted was shocked. Oley walked over to Ted's desk and looked into the pitcher.

"Where did you get my dad's pellets?" he asked Haywood. "Dad asked me to give those pellets to the president when I received the Congressional Medal of Honor."

"Your dad?" Haywood said. "Ted, now do you understand why Oley needs security protection until he returns to the Air Force?"

Ted nodded. "Are you telling me that these pellets are responsible for turning water into gasoline?"

"You tell me," Haywood said, motioning to the pitcher of gasoline. "Oley, how is your dad involved?"

"My father was a military scientist who was working on a top secret blue book project. He told me these pellets were from some friendly people from another world and asked me to give the pellets to our president. He said he was afraid of what could happen if they were released to the public."

Haywood cleared his throat. "Captain, due to your amnesia, you gave these pellets to several different people who were kind to you. Most of them believed they were candy that you found in their dumpsters and they threw them away or on the ground. But when it rained, the pellets turned the rain water into gasoline."

Oley looked surprised. "I gave these pellets to people? If my father finds out, I'm in deep trouble." Haywood looked at Ted Ruffen and then shook his head. "I'm afraid I have some bad news for you, Captain. Your father passed away. He died by suicide over a decade ago after you disappeared from the hospital."

Oley sat down in shock. He looked at Ted with tears in his eyes. "Suicide?

That can't be . . ."

Ted handed Oley an open FBI file with for Oley to read. Oley read the first page, then shook his head. "No way my father would leave my mother by killing himself. They say here that he left his car running in the garage. But somebody must have put him there. It had to be someone who knew my father was working on the blue book project. Where's my mom?"

"Do you know who Air Force General Mike Golden is?" Ruffen asked.

Oley looked at Haywood and Lilly who were sitting together on the office sofa. "I've heard his name mentioned. Why?"

"The general told me he had talked with your mother and she's well. General Golden told us to keep you here until he sees you and can get you back into the Air Force where they can protect you. You can stay here in my office until then. That sofa converts into a bed, and my men will bring you meals and protect you until you return to the Air Force."

Oley nodded. "Do you still have my dad's pellets?" he asked Haywood. "I have them and I will return them to you soon as I can."

"Please be careful who you tell about them," Oley said. "Don't give them away to anyone."

HAYWOOD'S SURPRISE

AYWOOD DECIDED TO SWING by the coffee shop in Detroit where he first met Lilly years ago. Noticing the sign, Lilly said, "Honey, look over there between the stores. It's the coffee shop where we met."

"Let's park the car and go in for some coffee and those great beef sandwiches," Haywood suggested.

They asked to be seated at the same table where they had their first date.

Haywood had just ordered their meal when his cell phone rang.

"It's Ruffen," Haywood told Lilly after reading the caller ID. "Hi Ted.

What's up?"

"I've received a call from Kyle Hampter. He told me he would like for you to be back at my office when he arrives here in about an hour. You can bring Lilly as well."

"Okay," Haywood said. "We'll finish eating first and then head back."

Lilly overheard their conversation and smiled. "It sounds like Ted has something important to tell us. I hope it will be great news for Oley."

When Haywood and Lilly returned to Ruffen's office, Ted was with Oley and Kyle Hampter. Hampter was telling them about his meeting with Lester King. After introductions, Haywood and Lilly found chairs and sat down to listen.

Kyle pitched a small envelope from his coat pocket onto Ted's desk. Ted opened it and whistled, showing everyone the money inside.

"This is what I received from Lester King," Hampter said. "He told me 'more to come with good information.' King believes he can get the information from me about who has the formula to turn water into gasoline. I saw two of Lester King's bodyguards, who were armed. I was warned not to tell anyone about our meeting, as he or his men would come looking for me."

"That's a threat," Oley said. "I'll bet they're the ones responsible for my father's death!" Ted handed the envelope back to Kyle.

"I don't want the money," Kyle said. "I thought you'd want to hear about this."

Ted patted Kyle on his shoulder. "Would you be willing to do some extra FBI work? Haywood, you've been sitting there quietly. I'd love to hear your comments and have your help as well."

Haywood looked at Lilly. "Sweetheart, I know I'm supposed to be retired, but Kyle and Oley are going to need my help."

"Of course, honey," Lilly said, "if it helps Oley and Kyle, but please be careful!"

Haywood nodded. "What do you want me to do, Ted?"

"When is your next meeting with Lester King?" Ted asked Kyle.

"I have to wait for his call," Kyle said. "I'll notify you then. He likes very expensive meals, so we'll probably meet at a restaurant."

Ted raised the envelope.

"Give it to the needy," Kyle said. "I don't want it."

"Okay, Haywood, let's make plans for the next meeting," Ted said.

"I want to know who murdered my father because he would never have committed suicide," Oley said. "I called my mother after our initial meeting this morning and she agrees with me. It had to be murder."

Haywood nodded. "Hopefully we'll discover information about what happened to your father from our Lester King investigation."

NEITHER Haywood nor Lilly spoke much during the ride home. Country music was playing over the radio, and when the music stopped, a news report came on about the rising price of gasoline due to the possibility of a hurricane on the East Coast.

When they arrived home, Haywood opened the garage door. To their shock, the door to their house from the garage was ajar and several items were strewn all over the garage floor. Haywood ducked into the car, opened the glove compartment, and removed his .45-caliber revolver.

"Stay in the car," he told Lilly. "If you hear gunshots, call the police." "Let's call the police now," Lilly said. Haywood pitched his cell phone into Lilly's lap and entered their home with his gun out. The place was a mess, with shelves overturned and papers and knickknacks scattered all over the carpet. It was apparent that whoever broke in was looking for something other than money. The desk in their office was still standing, but the drawers were removed and stacked on the floor. However, the cash inside Lilly's budget book was not touched, nor was his checkbook. Satisfied there was no one in the house, Haywood returned to Lilly who was still sitting in the car talking to the police dispatcher. The police arrived a few minutes later.

"Is this the Haywood Runyan residence?" the officer asked. "Our chief said you retired from the FBI. Is that right? A woman called from this address about a thief in the home."

Haywood got out of the car and shook the officer's hand. "That's right," he said. "I already checked inside and found no one. They left a mess, though. It's obvious they were looking for something and couldn't find it. It doesn't appear they stole any money."

After filling out a report, the officer left. Lilly walked in the house and was shocked to see the mess. Haywood suspected what they were after and checked the backyard by the bushes. Nothing had been disturbed. The pellets were still safe.

The next day Haywood traveled to Detroit, with pellets popping and humming on the back seat and in the car trunk. After parking, he threw a jacket over the pellets so that they couldn't be seen through the car windows. Ted was sitting at his desk and Oley was laying on the sofa as he walked into the office.

Oley jumped up when he saw Haywood. "Did you bring my father's pellets?"

"Yes," Haywood said, smiling. "They're in my car."

"I suggest using the underground parking lot so no one sees you," Ted said. "I'll call and make sure you have clearance."

Haywood and Oley left Ted's office and drove to the underground parking lot. Hearing the noise from the pellets, Oley looked under the jacket and removed the pellets.

"Thanks for bringing these to me," he said. Haywood nodded. "There's more in the trunk."

Once parked, Haywood opened the trunk and both were amazed at how they were popping, humming, and multiplying. "These pellets are really producing," Haywood said. "It's amazing to watch them."

"Why don't you keep the pellets in your trunk," Oley said. "It'll be my way to say thank you to you and your wife, my little angel, who protected me from the protesters."

Oley returned to Ted's office with his string of pellets while Haywood parked back in the above-ground parking lot.

As Haywood returned to Ted's office, Ted and several FBI agents were looking at the string of pellets. Then Oley hid the pellets under the sofa and left with the agents.

When left alone, Haywood told Ted about their house break-in.

"I didn't find anyone in the house and they didn't find the pellets," he said. "I'm glad I had the foresight to bury them in the backyard before I left."

Ted looked thoughtful. "Maybe Oley's right about Lester King and his bodyguards murdering his father."

"Where did Oley go with your agents?" Haywood asked.

Ted laughed. "Apparently Oley had a craving for a hamburger and fries."

Haywood smiled. "That does sound delicious. I should go home so I'm not late for dinner."

Ted's secretary opened the office door. "Kyle Hampter is on the line," she said.

Ted picked up the phone. "This is Ted."

"I just talked to Lester King," Hampter said on the other end. "He wants to meet at Bob Plum's Restaurant tonight at 7 p.m. He's flying here from his East Florida office. The restaurant's close to the Detroit airport. Are we still following the plan we discussed at our meeting?"

Ted motioned to Haywood. "Kyle Hampter is going to meet with Lester King tonight."

Haywood nodded. "I need to get home, but let me know if you need me as backup." Lester King's bodyguards, Monty and Tolly, flew into Detroit early and gave fraudulent credentials to the car rental agent. They were soon driving a new Cadillac to St. Treba Catholic Church. King's instructions were to grab Father Pario and hold him as extra security for the 7 p.m. meeting with Kyle Hampter. When Monty saw the Dilly's Pawn Shop he pulled over.

"We can trade our old guns for new guns here," he said.

As they walked into Dilly's Pawn Shop, a large man from behind the counter greeted Monty with a handshake. "What the hell you doing here, Monty?"

"Hey Dilly," Monty said. "This is Tolly. We'd like to trade our old guns for some new ones. Do you have any?" Dilly nodded. "But I can't sell you any guns until the state government checks your background."

Monty reached into his pants pocket and set down two thousand-dollar bills. Dilly looked them over, then placed the bills in his pants pocket and pointed to the new gun rack.

"Give me your gun," he said to Tolly, who handed Monty his loaded gun, an older .40-caliber revolver. It was replaced with a newly loaded .40-caliber revolver. Monty gave the bullets he had removed from Tolly's revolver back to him. Then he put his older .45-caliber revolver on the store counter and grabbed a new gun for himself. Monty removed the bullets from his old gun. He replaced the bullets with new ones, leaving the old ones on the counter. Then he placed the new gun in his hip holster.

Dilly was looking curiously at Monty. "Where'd you get all the money?"

Tolly laughed, and Monty smirked, replying, "Hey, Dilly, that's none of your business."

"Yeah," Dilly said. "I just never seen you with that much money, and I don't want to see you go back to prison."

Monty shrugged. "If you must know, Tolly and I work for Mr. Lester King, who owns West-East Coast Oil Global Industry."

Dilly smiled, "Sorry, Mont, but when you threw the bills at me, I thought you and Tolly may have robbed a bank, and I don't want the local police in my shop."

Dilly removed the old guns from his counter. Monty and Tolly headed for St. Treba Catholic Church to pick up Father Pario under Lester King's instructions. They parked the car near the entrance and waited. When they saw the priest walking out of the church, they left their vehicle and grabbed the priest. Tolly showed him his revolver in his shoulder holster.

"Don't give us any trouble," Monty said, "or we'll have to shoot you." Tolly shoved the priest into the back seat of the Cadillac and joined him.

Just then, another priest came out of the church with a group of children who had been attending a youth fellowship class.

"Father Pario, Father Pario," two of the girls shouted. "Two men just forced Cardinal Blessings into that new Cadillac in the church parking lot!"

"Cardinal Blessings? What's going on?" one boy shouted. He and several other boys ran towards the Cadillac.

Cardinal Blessings turned to his assailants. "My sons, what's going on here?"

"Father, be quiet," Tolly whispered.

Monty looked at the priest in his back seat. "Are you Father Pario?" Cardinal Blessings pointed at the other priest who was with the children.

"I'm Cardinal Blessings. Father Pario is over there with the children."

"Tolly, we got the wrong guy!" Monty hollered. "Get him out of this car now, and let's get out of here."

Tolly opened the car door and pushed Cardinal Blessings from the back seat into the hands of the boys standing by the car's rear door. Monty drove the car out of the church parking lot, almost hitting the children, and Father Pario ran to Cardinal Blessings.

"Are you okay, Your Eminence?" Father Pario asked. "What happened?"

Cardinal Blessings stood, a bit shaken. "Those two thugs in that Cadillac thought I was you. Apparently, they were trying to kidnap you."

Father Pario looked confused. "Did they say why?" Cardinal Blessings shook his head.

"I got the plate number!" one of the boys shouted, handing a piece of paper to Father Pario.

Father Pario called the police on his cell phone, giving him the license number and explaining the kidnapping attempt.

MONTY and Tolly pulled into a large mall parking lot, leaving the car's engine on. "Lester King is going to be very unhappy with us for mistaking Cardinal Blessings for Father Pario," he told Tolly. "This could mean a death warrant for both of us. I'm outa here." With that he left the vehicle and headed for a store in the mall.

Tolly jumped into the car driver's side and drove away. A policeman spotted the vehicle and several police units began following Tolly with sirens blaring and red and blue lights flashing. Tolly pulled his .40-caliber revolver from his holster, stopped the car, and shot at the police, who returned fire. Then he jumped from the car and fell to the road, but not before he was hit by several bullets. An ambulance rushed Tolly to the hospital, where he was placed in surgery and returned to a room with police security.

Tolly was unconscious, but kept repeating, "Father Pario, Father Pario, Father Pario." One attending nurse knew Father Pario and was a member of St. Treba Church, so she called Father Pario, telling him about Tolly Sea- more, who was calling out his name and would likely die from his injuries.

Father Pario drove to the hospital. The doctor told him that the bullets had punctured Tolly's lung and he would not survive. Father Pario approached Tolly's bed, leaning over him. "Son, should I talk to God to save you from your sins?"

Tolly could hardly talk, but nodded his head. Father Pario held his gold cross and began praying and asking forgiveness for Tolly. "What do you need our wonderful God in heaven to forgive you for?" he asked.

Tolly whispered, "Murder. Please forgive me, God. I have been doing terrible wrongs for Lester King. Monty and I murdered a guy in his garage some time ago by placing him in his car and leaving the car engine on and closing the garage doors."

A police officer was standing near the priest and began jotting down what Tolly said.

"Son, who was the person you and Monty murdered for Lester King?" Father Pario asked.

"He was living in Florida. His name was Oley Washington, and he had a big secret to ruin Lester King's oil industry. Oh God, I am so sorry, please forgive me."

"What happened to Monty?" the police officer said.

Tolly was close to death and could only talk in a whisper. "A mall" was all he could say. Those were his last words. The doctor arrived and pronounced him dead, placing the sheet over his head. Father Pario gave Tolly his last rites and asked God to forgive Tolly for all of his sins.

The police officer who took the notes called the police dispatcher, reporting that Monty Hellman was in the Detroit Mall and was wanted for kidnapping and murder.

"He may be armed and dangerous," the officer warned.

From there, he contacted the Detroit police chief, explaining what Tolly had said before dying. The chief contacted Ted Ruffen.

Because of Monty's prison record, the Detroit Police were able to locate his mug shot. Several officers entered the main Detroit mall and spotted Monty sitting next to a woman with children. He appeared to be watching them play and was smiling and talking to the pretty woman when police surrounded him with their guns drawn.

"Stand up very slowly, Monty, and keep your hands up," one of the officers said. Monty's revolver was removed from his hip holster and he was told to lie face down on the mall floor. Then they handcuffed him and pulled him back to his feet.

"I want my attorney," Monty said.

"You're gunna need a very good attorney, as now you are being charged with kidnapping and murder," one of the officers told him.

The woman who was sitting next to Monty grabbed her children from the swings and left the mall in a hurry.

HAYWOOD was home helping Lilly clean up the mess left by someone who was looking for the pellets in their home when the phone rang.

"Your former boss is on the phone," Lilly said. "Are you sitting down?" Ted asked Haywood. "I'm sitting," Haywood said, still standing.

Lilly watched as Haywood slowly sat down as he was getting the news. "Tolly told Father Pario that Lester King gave orders to murder Oley Washington Sr. in Florida. So it was not suicide," Ted said. "Lester King is flying into the Detroit Airport to meet with Kyle Hampter. You know we had made plans for that meeting, but now it's not necessary, as my agents are waiting at the airport to arrest Lester King for accessory to murder and kidnapping. And no doubt they broke into your home."

"How is Oley handling this news?" Haywood asked.

"He's not shocked, but as you can imagine, he's pretty angry. At the same time, I think he's relieved that the truth about how his father really died was discovered. He's been sharing the news with his mother. General Golden will be here tomorrow to take Oley with him back to the base in Nevada, as the colonel there is retiring. General Golden told me he talked with the president, who wants to meet the captain and place the Congressional Medal of Honor around Oley's neck."

"That's great, Ted," Haywood said. "Please tell Captain Oley Washington Jr. that Lilly and I wish him and his mother the very best."

After hanging up, Haywood gave Lilly a hug and kiss, telling her about the news. He also told her about the pellets that Oley had given them.

"Now I have to start writing my book about Captain Oley Washington Jr.," he said.

"It'll be fantastic," Lilly said, embracing her husband.

"As long as we continue to have water here in our home, we will never have to buy gasoline again."

"Thank you, Captain Oley Washington Jr.," Lilly said.

FATHER PARIO
RECEIVES A GIFT

Several days later Haywood asked Lilly, "Sweetheart, I wonder if Oley Washington Jr. has left Detroit with General Golden?"

Before Lilly could answer, their phone rang. Lilly smiled at Haywood. "Maybe that's Ted with an answer to your question."

Haywood nodded and picked up the phone. "Hello? Ted, yes I am sitting down," he said, still standing, and pointed to Lilly and the phone for her to listen. Lilly went to the other room, picking up the phone to hear Ted talk.

"Haywood, I received a call from Captain Oley Washington Jr. He will be in at the capital to receive the Congressional Medal of Honor from our president. I've talked with Father Pario, and he is attending with Father Steven and Sister Martha. We will fly to DC on our federal plane, and I hope you and Lilly will join us. We will leave the Detroit Airport Tuesday next week at 11:00 a.m. I'd love it if you can come and also meet the FBI director."

Haywood looked in the other room at Lilly, who was nodding. "Thanks, Ted. We would love to come."

Tuesday arrived, and Haywood and Lilly greeted Father Pario, Father Steven, Sister Martha, and Agent Ted Ruffen on the federal plane.

Father Pario stood as all took their seats and offered a prayer to God for a safe journey.

When the plane landed at the Washington, DC airport, the FBI director was waiting at the bottom of the steps. He greeted everyone with handshakes and hugged Lilly and Sister Martha. The director was rough looking man, younger than Ted Ruffen, with a medium build and wavy brown hair. He spoke with a Brooklyn accent and wore a gray polyester suit, white shirt, and a golden-colored tie held in place by a set of gold handcuffs. An American flag was pinned to his lapel. Placing his hand on Haywood's shoulder, he smiled and said, "Haywood, we miss you since you retired. But now that I see your wife, Lilly, I understand why you've retired." When they arrived at the White House Rose Garden, they were escorted to the very first row of seats. The media was also arriving, and everywhere you looked were beautiful flowers.

The president and vice president arrived with several Secret Service agents. The president walked to the podium and stood between the American and president's flags blowing in the wind.

"My friends, we are gathered here in our Rose Garden to honor Captain Oley Washington Jr., who helped American soldiers and himself escape from a Vietnam military prison during the Vietnam War. He was to receive this honor many years ago, but was beaten up by a group of war protesters and placed in a hospital. He later disappeared with amnesia. Please welcome Captain Oley Washington Jr."

The White House doors opened, and Captain Oley Washington Jr. with his mother holding his arm, approached the president together. Lilly Runyan stood and clapped her hands, and the crowd followed her. The president requested all to be seated and placed the medal around Captain Washington's neck as all continued clapping. Then the president requested General Mike Golden to join them on stage. Captain Washington stood at attention and saluted the general, who returned the salute as the president started to speak. "Captain Oley Washington Jr. has been a captain in the United States Air Force for many years, and now I will let General Mike Golden speak."

The general looked directly at Captain Washington, who was still standing at attention. "At ease, Captain. Thank you, Mr. President. As of today you are no longer a captain in our Air Force, but are promoted to a colonel. Oley Washington Jr., you will now be in charge of the US Air Force base in Gladstone, Nevada where Colonel Harrington just retired."

Once again Lilly stood and clapped as all in the Rose Garden joined her. General Golden requested all to be seated as he removed the captain bars and replaced them with new colonel medals.

Colonel Washington received a kiss from his mother, who told him, "Oley, your dad would be so proud, and I know he is looking down from heaven and smiling."

After the ceremony, the president invited all into the White House Dining Room for lunch. But before leaving the podium, he pointed to Lilly Runyan, saying, "Lilly Runyan, I have a Citizen's Medal for you, and you certainly deserve one for saving this colonel many years ago from the war protesters who beat him up and injured him here in Washington, DC. Please step forward and receive this medal."

The president placed the medal around Lilly's neck and thanked her while the crowd cheered. Colonel Washington hugged Lilly and thanked her. Colonel Washington's mother, walked over to Lilly and gave her a hug. All were standing again and clapping as the president continued speaking. Next he pointed to Haywood Runyan, then Father Pario, Father Steven, and Sister Martha. "All of these people are responsible for saving Colonel Oley Washington Jr. Haywood Runyan is writing a book about the adventures of the colonel while suffering from amnesia in Detroit. Haywood, I and the First Lady would love to have a signed copy of your book, and the vice president whispered to me he would like to have a copy for his family. Okay, everyone, let's go into the dining hall and have some lunch."

After being seated in the White House Dinning Hall, Haywood was approached by several people, including General Golden, asking him where they could buy the book. Lilly looked at her husband and smiled. "I believe you're going to be very busy writing your book when we get home." Haywood smiled, and nodded his head in disbelief.

Ted Ruffen approached Haywood. "You know what I want from you, don't you? And the director wants one too."

Haywood laughed. "Okay, okay, but first I need to write it."

Colonel Washington sat down next to Father Pario, removed a signed check from his pocket and handed it to Father Pario. "That's my government check from the Air Force with my back pay for the years when I was missing," he said. "I no longer need the money, and, Father Pario, my mother and I would love for you to use this money to provide a children's recreation center in my father's name. We'd love to buy the property of the old bank where I used to live and have the new children's recreation center located there. Can you do this?"

Father Pario stood up from the table, showing everyone the check from Colonel Washington and telling all what he had been told by the colonel and then offered a prayer to God for all those attending, their safety, a blessing for the colonel and his mother, and a children's recreation center to soon be built with thanks to Colonel Washington and his mother. Lilly stood up from the table, clapping her hands with the president, First Lady, and vice president with his wife, followed by the crowd. When Lilly sat down, Haywood gave her a hug and kiss. "I'm a very lucky man having you as my wife," he said.

Lunch was over, and the group returned to the airport to fly back to Detroit. They all shook hands with the FBI director and said their goodbyes.

Haywood sat with Father Pario, Father Steven, and Ted Ruffen brainstorm ideas for the new children's recreational center in Detroit, while Lilly and Sister Martha talked about the ceremony and the Citizen Medal Lilly had received from the president.

KING HASSMEN TAVIO

KING HASSMEN TAVIO AND the Saudi Arabian Royal Oil Empire had seen the media coverage about Colonel Oley Washington Jr.'s ceremony and were now aware that the pellets turning water into gasoline were with the president of the United States of America. He wasn't happy that his agents were unable to find any pellets in the Runyan home.

"Now the president has the pellets," he told his fellow diplomats. "Why were my direct orders not followed? How did they get those pellets to the capitol? Why weren't they stopped?"

"King Tavio, " a very nervous diplomat replied, "I gave all your instructions to our agents. They broke into the Runyan home and found no pellets. The captain and the Runyans were at the FBI office in Detroit, Michigan. I have not been able to contact our agents for several days. We had information that Oley Washington Jr. was under federal protection and staying at the FBI office in Detroit. Do you have any new instructions for our agents?"

"My friend Lester King has been arrested for kidnapping and murder!" King Tavio hollered. "The United States government now has the pellets! Do you know what this will do to us? The US will stop buying our oil! Lester King was trying to stop this, and now we have no choice. Captain Oley Washington Jr. is now an Air Force colonel in charge of an Air Force base in Gladstone, Nevada. We can't get to him, so get the Runyans. Got it? When the agents get the Runyans, they are to contact me immediately, understood?"

"Yes, my king," the diplomat said.

HAYWOOD and Lilly had returned home and looked at the damage from the thieves who had broken into their home several days before.

"We'll have to go back to the casino to get the money to fix all this," Haywood said. "Maybe you can bring home lots of money to replace this damaged furniture." Lilly was too sad about the damage to laugh.

When the doorbell rang, Haywood grabbed his revolver and put it in his side pants pocket before slowly opening the door. A small package was left on their porch. The package had been delivered by the mailman, who was walking to his mail truck in their driveway. Haywood waved at the mailman, who waved back, then picked up the package, which was sent from the FBI office in Detroit. He took the package in the house, showing it to Lilly.

He slowly opened the package and found an envelope with a note from Oley Washington Jr.

"To my best friends," the letter read. "I will always be grateful to you both. Ted Ruffen told me about the damage done inside your home. I am sorry for what happened. I had a secret pocket inside my military grip. It's not a lot of money but may help you with the damage to your home. My mother and I will be looking forward to seeing you soon. You are part of our family now. Love you both, Oley Washington Jr."

Lilly reached for the envelope while Haywood was reading the note. "Honey," she said in tears, "there's almost twenty thousand dollars in this envelope."

KING Tavio was in his palace eating when he was interrupted by his diplomat. "You are interrupting my meal," he said abruptly. "This better be good or I will have your head."

The diplomat smiled. "King Tavio, what I am about to tell you will make you smile! Our intelligence agents told me that the United States president gave the pellets to his chief in the military to use for their military vehicles only and cannot be used by the public. The pellets will be secured by the military, and no one, including the American people, will have access to any pellets."

King Tavio got up from his golden chair and hugged the diplomat. "King Tavio," the diplomat said, "what do you want to do about the Runyans?"

"Nothing," King Tavio replied. "As long as the pellets remain only with the United States military we will continue to make money."

LILLY MEETS LORI, HER COLLEGE ROOMMATE

A T THE RUNYAN RESIDENCE, Lilly's cell phone began playing music. When she noticed the caller ID she quickly answered it. Her former college roommate, Lori Docker, was waiting to talk to her. "I don't believe this, Lori. I can't believe this is you," Lilly said.

Haywood was watching television when he heard Lilly talking on her phone. He was curious, so he approached Lilly, who whispered to him, "Honey, it's Lori Docker, my old college roommate."

"Oh, Lori, are you kidding me?" said Lilly over the phone. "You've never been married? You've been working all these years as a children's doctor and just transferred from Washington, DC, to Detroit? You're working at the Little Ones Hospital? This is so great! When can we get together? I have so much to tell you."

Haywood started walking away when he heard Lilly mention Oley Washington Jr.'s name. He stopped, turned around, and approached Lilly shaking his head. He thought that their home may have been bugged when it was broken into.

"Sweetheart," he whispered to Lilly, "please don't mention Oley's name, as some foreign agents may be listening."

Lilly nodded, telling Lori, "We need to meet together so I can tell you more about this person. Yes, Lori, we can meet at Pop's Restaurant, close to your hospital. It will be great seeing you again. I think the last time we saw each other was when you were maid of honor at my wedding. I can't wait to see you tomorrow at 12:30."

Lilly placed her cell phone in her pants pocket and looked at Haywood. "What a fantastic delight talking with Lori. Do you remember her?"

"Of course," Haywood said, smiling. He knew Lilly very well and wasn't surprised by her next statement. "Honey, wouldn't it be wonderful if Oley and Lori got together?" Haywood nodded.

"Honey, can we go to Detroit tomorrow and meet Lori for lunch? You take me to the restaurant and then you can visit Ted Ruffen or Father Pario."

Haywood nodded. He entered the garage and poured water into a pail and threw pellets into the water. He forgot that the garage had windows where someone could watch. He poured gasoline into his tank until it started overflowing, not realizing there was a hidden camera in his garage.

THE next day Haywood and Lilly entered Pop's Restaurant. Lilly pulled Haywood over to the table where Lori Docker was sitting. Lori was wearing her doctor's jacket and looking very attractive, younger than her age, and still maintaining her girlish figure. Lori stood up and hugged Lilly and Haywood. Haywood listened for awhile as Lilly told Lori all about Oley. Then he told them, "I'm going to visit Father Pario, Father Steven, and Sister Martha. It was great to see you again, Lori. Have a great time, and honey, call me when you want me to pick you up." He kissed Lilly and left the restaurant.

FATHER Pario was surprised to see Haywood, as were Father Steven and Sister Martha when he arrived at the church rectory.

"Haywood, come to my office with me," Father Pario said. "I want to show you our plans for the new Oley Washington Sr. Youth Center. You have no idea how thrilled the people are about a youth center in this neighborhood. No time is being wasted. I'll show you our building plans and then we will walk across the street."

Father Pario pointed to the building sketch posted on the office wall. The sketch showed a large, one-story building to replace the old vacant bank building. He chuckled while explaining the sketch to Haywood.

"We're going to keep Oley's huge safe in the new building for storage. This new building will have a full-scale gym and an Olympic-sized swimming pool with a tall diving board. There will be extra rooms with various games for children and adults." Pointing at the rear of the building, he explained, "The pool will be used for our swimming meets and for children to enjoy swimming." He pointed to another room. "This will be a chapel dedicated to Oley Washington Sr. and to his son, Colonel Oley Washington Jr. Let's go across the street where the gate was and into the woods. I want to show you what we are doing."

Haywood noticed many vehicles parked on the street. There was no longer a gate and people were busy removing vines and high grass on the fence. Several people were mowing the high grass with tractors. Engineers were measuring and applying stakes with flags. The old vacant brick bank building was gone.

"We've received approval from the Detroit city fathers and mayor to buy the land and build the new youth center. Many of these people working here are church members. It's a miracle, Haywood. Some people working here are volunteers from this neighborhood. This is happening because you decided to write a story about a person living in our neighborhood."

Haywood nodded in wonder. "Is the safe still here?" he asked.

Father Pario nodded. "Yes. It's so large we're building the new building around it."

They laughed and returned to the church. Father Pario checked the church mailbox, finding several letters addressed to him. There was no return address on the envelopes. Haywood watched as he sat at his desk, opening the envelopes and removing certified bank checks payable to the Oley Washington Sr. Youth Center for $100,000. Each envelope contained a certified bank check of the same amount and included a note. Father Pario begin chanting, "It's a miracle." He handed the checks and notes to Haywood.

The notes were all the same.

Father Pario,

Several of us during our college days are responsible for putting Oley Washington Jr. through hell. We are hoping you will pray for God to forgive us. We also have prayed to God asking for his forgiveness. We were a group of young college students listening to our media, as the American people wanted someone to stop the Vietnam War. We were foolish college kids. We didn't know Oley Washington Jr. was in Washington, DC to receive the Congressional Medal of Honor from our president. We hope Oley Washington Jr. will forgive us. Please use these checks to pay for the Oley Washington Sr. Youth Center. Thank you. "God has responded by entering their hearts, prompting their support of our new youth center," Father Pario said as Haywood returned the note.

Haywood nodded. The desk phone started ringing, and Father Pario answered it. Ted Ruffen's voice was on the other end. "Father Pario, is Haywood Runyan there with you?"

"It's Agent Ruffen," he said, handing the phone to Haywood. "Ted, this is Haywood. What can I do for you?"

"We just received information from our director in Washington, DC. Several foreign intelligence officers are somewhere in this country. We believe they were responsible for the damage in your home when looking for the pellets. I called your home and didn't get an answer. It paid off calling Father Pario. You and Lilly be careful because they might return to your home."

"I hope they do, as it will be payback time for what they did," Haywood said.

"Haywood I'm very concerned about you and Lilly. I talked to our director about the damage to your home, and I want protection for you two. The director has allowed me to give you two agents from our department. Agents Bicker and Bullord have volunteered. They are now at your local police department, talking with the police chief. They will be arriving at your home when you return. Say hello to Lilly for me."

"She's at Pop's Restaurant with a friend having lunch."

"Get over there," Ruffen said. "I'm sending two agents there now. Isn't that the restaurant near the children's hospital?"

"That's the one," Haywood said. "Lilly's friend is a Doctor Lori Docker, her old college roommate. I am leaving here now."

Haywood handed the phone to Father Pario and ran from the church to his car.

LILLY and Lori had finished eating their lunch and were sitting at the table talking about their college days and Colonel Oley Washington Jr. Two foreign men walked to a booth across from the table where the women were sitting. A waiter approached the men, and they ordered coffee. Lilly was becoming uneasy while the men talked and continued to look at them. One of the men stood and approached Lilly and Lori's table. He was young man, tall with a medium build and long black hair combed back in a ponytail. He was wearing dark slacks and a sport shirt displaying the University of Washington. He began speaking in heavily accented English, introducing himself. "Hello, girls, I am Abdue Palio, and my brother in the booth is Rabi." We heard you discussing the University in Washington, DC, where we are both students." Lilly smiled, introducing herself and Doctor Lori Docker. Lori then invited Abdue and his brother Rabi to sit at their table.

A waiter approached asking, "Is there a Lilly Runyan here? Mrs. Runyan, someone wants to talk with you on our phone." He pointed to the phone, and Lilly picked it up, hearing Ted Ruffen.

"Lilly, this is FBI Agent Ted Ruffen."

Lilly, concerned, asked "Is Haywood okay?"

Ted replied, "Lilly, I just talked to Haywood in Father Pario's office. Haywood is fine and on his way to Pop's Restaurant, and I am sending agents there as protection for you and your friend. You may be in danger. Don't meet or talk to any strangers. Haywood can explain to you why we are concerned."

"Thank you, Ted."

Lori was looking toward Lilly while the two men were talking, telling jokes about Washington University, and what it's been like staying there. Lilly motioned for Lori, who excused herself and disappeared into the women's restroom with Lilly.

After they left, Abdue whispered to Rabi, "Something is wrong. The one lady received a phone call and motioned for the other lady to join her. Both women went into the women's restroom. I think we need to leave this restaurant now." They didn't pay or tip the waiter and left the restaurant, running to their car.

SECURITY FOR LILLY, LORI,
AND HAYWOOD RUNYAN

ILLY TOLD LORI ABOUT the phone call from Ruffen. Lori, looked concerned. "Do you believe the men who were sitting with us were after us?" she asked. "Oh my! I mentioned Colonel Washington's name, and they both laughed when I left the table."

Lilly shook her head. "I don't know, but I was told for us to stay here until Haywood or the FBI arrives."

Haywood parked his car, noticing two unmarked FBI vehicles as he raced into the restaurant and looked for Lilly and Lori. They were not at the table. He approached a waiter, who pointed at the restroom door. A waitress walked into the restroom, telling Lilly that Haywood and several FBI agents were now in the restaurant.

Lori looked at her watch and gave Lilly a hug. "I have to return to my office."

They left the restroom, and Lilly walked into Haywood's open arms, receiving a kiss. Lori hugged both and left with an agent following her to her car.

The two strange men were gone. Lilly left Haywood to check on Lori, and the agent who walked Lori to her car reassured Lilly that Lori was okay.

Lilly thanked the agents, then said to Haywood, "There were two foreign brothers going to our university at Washington, DC. One was named Abdue Palio, and the other was Rabbi. Abdue approached us, and Lori requested them both to sit at our table."

A waiter came over to Haywood and Lilly. "When Lilly and her friend went into the restroom, the two foreigners left and didn't pay their bills," he said.

Haywood nodded. "Sweetheart, it's a long way from here to Washington, DC."

"Lori and I thought they were young students. We wanted to talk with them about their college experience," she explained. "I'm sorry."

An agent approached Haywood. "Head Agent Ted Ruffen wants us to follow you and Lilly home," he said.

"Thanks, but it's not necessary for you to follow us," Haywood said.

The agent laughed. "That's his orders. You know. You used to work for him. Do you want to call him?"

Haywood, laughed, shaking his head. "You made your point. Let's go home."

THE next day, Haywood's phone rang. When Haywood answered, Colonel Washington's voice was on the other end of the line. "Haywood, how are you and Lilly? I'm all settled in my new position here in Nevada. I've taken command of the air force base. It's been quite an experience. I've recently received new stealth bombers and pilots, and we are involved in training. I was wondering if you got my envelope that I sent you with money to repair the damage to your home. Agent Ruffen said you had a lot of damage."

"Yes we did, Oley. Thanks!" Haywood said. "That's a lot of money, and we have insurance to cover some of the damages. So I'll just keep this money to give back to you when we see you again."

"I don't need the money," Oley said. "Keep it for you and Lilly. Are you two doing okay?"

"We have FBI agents staying here with us," Haywood explained. "Apparently, King Tavio, owner of the Saudi oil cartel, has his intelligence officers looking for us." Haywood talked loudly so if any hidden listening devices were in their home, they could hear him. "King

Tavio is a friend of Lester King, who was involved in your father's death. Lilly wants to talk to you."

Lilly took the phone from Haywood. "Colonel, there is someone I would like you to meet. She's a children's doctor in the Little Ones Hospital in Detroit. She was my roommate in Washington, DC, when we were going to the university. She has never been married and is very attractive, and I have told her all about you. When you come to Detroit for your dad's youth center opening ceremony with Father Pario, I'll introduce you to her. Her name is Lori Docker."

"I don't know what to say except thanks for all you have done for me," Oley said. I'll be at the opening ceremony with my mother and look forward to meeting Lori and seeing you and Haywood, Father Pario—all my new friends."

Lilly returned the phone to Haywood. "Oley, It's me," Haywood said.

"I talked with Kyle Hampter yesterday," Oley said. "He told me the government has classified the pellets as 'Can Do.' Top-secret and military labs are working on the formula and will use Can Do for their fuel. Haywood, that was not my father's plan—"

Haywood interrupted. "Don't worry. When the military gets all the gasoline they need, they should declassify the pellets and make them available to the public."

"I hope you're right. If my father were still alive, he would be very upset about what the military is doing with the pellets. The price per gallon in Nevada is now six dollars or more. Well, better go. It'll be great to see you and Lilly at my father's youth center opening ceremony."

Haywood hung up the phone, shaking his head and looking very disturbed. "What's going on?" Lilly asked.

"Oley told me the pellets are now top secret," he whispered. "They're called 'Can Do,' and the military labs are working on the formula. He's not happy because his father wanted the president to give the pellets to the public."

Because all the furniture had been destroyed in the break-in, Agents Bullord and Bicker were sitting on the floor in Runyan's home while Lilly fixed meals for them.

When they were finished eating, Haywood announced, "Lilly and I are going to the Detroit Mall to get new furniture."

"We'll go with you," said Agent Bullord.

"Then who will protect our home?" Haywood asked.

The agents agreed to stay at their house while Haywood and Lilly left to go shopping.

KING HASSMEN TAVIO
TRIES ONE MORE TIME

ONE OF THE KING'S diplomats in the United States called King Hassmen Tavio.

"King Tavio, your intelligence officers who broke into Haywood Runyan's home looking for pellets planted several listening devices in their home. The officers were at Pop's Restaurant in Detroit, Michigan, yesterday making contact with a woman, Lilly, who is Haywood's wife, and she was with a friend who was dressed like a doctor. Lilly received a phone call and she and her friend went into the women's restroom and our officers left. They heard today on their listening devices hid in the home that Haywood and his wife were going to the Detroit Mall."

"Didn't one of my people tell you to leave the Runyans alone unless they still have pellets that could be given out to the public?" King Tavio asked.

The diplomat replied, "Yes, your orders were told to our intelligence officers, and they still believe the Runyans have pellets. They have seen from a hidden camera in the Runyan's garage yesterday that Haywood was using pellets and water to make gasoline and put it in his car's gas tank."

"We need to get those pellets," King Tavio said.

"With your permission, our officers will get back in the Runyan home to find the pellets today while they are at the mall," the diplomat said. "Do we have your permission?"

King Tavio paused several minutes while the diplomat waited. "Okay," he answered slowly. "Find the pellets and hold them for me."

FBI Agents Bullord and Bicker were sitting on the floor in the Runyan home watching television when they heard the sound of a window breaking in the garage. They jumped up, drew their guns, and pushed open the door to the garage.

Abdue and Rabbi were shocked when they saw two guns pointed at them. Agents Bullord and Bicker placed both men on the floor, handcuffing their hands behind their backs. Agent Bullord told Abdue and Rabbi that he was going to announce their capture to the media unless they showed the agents where they hid the listening devices in the house.

"If we remove the listening devices, will you let us go?" Abdue asked.

Agents Bullord and Bicker nodded, and Abdue began removing the listening devices.

"This is the last one," Abdue said, pulling out a knob from the oven. It looked exactly like the other knobs on the oven but was a listening device.

Altogether, there were six listening devices given to Agent Bicker. "You have them all now," Abdue said. "Can we leave?"

"This is the second time you and your friend have broken into this house because you knew where you placed the listening devices the first time you were here," Agent Bullord said. "You have done much damage to this home; therefore, you and your friend are under arrest."

Agent Bicker contacted the local chief of police, who sent three patrol cars to the Runyan home. He placed the men under arrest and rode with them to the jail.

KING Tavio was having a hard time contacting his diplomat in the United States. He was trying to find out if the pellets were found, but he had to call several times before he was finally able to reach the diplomat.

"Did our officers find the pellets?" he asked.

"King Tavio, you need to sit down," the diplomat said. This time King Tavio shouted, "Do we have the pellets?"

"No. Our men have been arrested and are now in the county jail awaiting trial," the diplomat answered.

"How did this happen?" the king demanded.

"When our officers broke into the Runyan home, two FBI agents were waiting for them," the diplomat told him. "They tricked them into removing the hidden listening devices, which confirmed to the agents that they were the men responsible for planting the devices when they originally broke into the house looking for the pellets."

There was a long pause on the other end of the line. "King Tavio, are you still there?" the diplomat asked.

"Yes I am here," King Tavio said. "I don't understand how this could have happened. Weren't the listening devices turned on? Why didn't they know someone was in the house?"

"They heard Haywood talking about going to the Detroit Mall to buy furniture," the diplomat explained. "They heard a television, but thought it was left on. When they saw Haywood and his wife leaving the house, they thought it was empty and attempted to enter it. But two FBI agents were still in the house, and our officers were arrested. What do you want to do now?"

"Our officers need to be returned to their home country, so get a great attorney," King Tavio said. "Do you understand my orders?"

"Yes, my king. I understand."

AGENT Bullord contacted lead Agent Ted Ruffen, telling him about the arrest and that Haywood and Lilly were at the Detroit Mall looking for furniture.

Agent Ruffen told Agent Bullord, "You and Agent Bicker stay with Runyans until I contact you. You both made a great arrest. Thanks."

Haywood and Lilly were in the Detroit Furniture Store looking at sofas and chairs to replace the ones damaged by the foreign officers when Haywood's cell phone rang. "It's Ted Ruffen," he whispered to Lilly.

"Hello, Ted," he said. "Lilly and I are shopping. What's up?"

He listened to Ted explain about the arrests. Then he relayed the information to Lilly. "Too bad I wasn't there," he told Ruffen. "It would have been payback time."

"Haywood, the men arrested were working for King Hassmen Tavio and were released from the county jail and sent back to their own country. Our director is satisfied that Agents Bullord and Bicker can now return to Detroit."

"Ted, Lilly wants to talk to you," Haywood said, handing Lilly his phone. "Agent Ruffen, thanks for calling me at the restaurant and contacting Haywood plus sending your agents to protect me and Lori," Lilly said.

Ted chuckled. "Okay, Lilly. I'm glad that you and Haywood are safe now.

You take good care of that retired FBI agent."

"You bet I will," Lilly said. She handed the cell phone back to Haywood, who also thanked Ted for protecting them.

KING Hassmen Tavio had his US diplomat on the phone again. "Bring our two officers back to me soon as possible, and you come with them. Do you understand?"

"King Tavio, what is going to happen to the two Intelligence officers and me when we return?"

"You just do what I say. Understood?" "Yes, my king."

Fall, winter, and early spring passed by at the Runyan home. Haywood and Lilly were no longer concerned about their safety, and they purchased new furniture, china, mattresses, frames for pictures, and several new figurines. Haywood fixed the garage back door, which had been forced open, replaced the window in the door leading from the house into the garage, and placed new locks on both doors.

After Haywood finished cleaning up the garage, Lilly had one more request. "Since everything has now been fixed and replaced, I hope you will get rid of those pellets. You've got a new desk and now you can write your story, but please get rid of those pellets."

Haywood started to reply when the phone rang. It was Father Pario. "Haywood, the Oley Washington Sr. Youth Center is finished. The grand opening ceremony will be on Saturday. Sister Martha has talked to Colonel Oley Washington Jr., who is planning on bringing his mother and a friend. Will you and Lilly be attending the ceremony? We would love to see you."

"We will be there," Haywood said. "Lilly will bring Doctor Lori Docker to meet the colonel."

"It's a miracle! It's a miracle!" the priest exclaimed. Haywood smiled and hung up the phone.

OPENING CELEBRATION

THAT SATURDAY, HAYWOOD DROVE their car into a newly paved parking lot. For a time, he and Lilly peered out the windshield admiring the new building. Many adults and children were walking toward the entrance. A huge banner was hanging next to the open doors, declaring "Grand Opening Ceremony." The building housing the Oley Washington Sr. Youth Center was made of limestone and was much larger than the old bank building that used to stand in its place. Instead of being covered by overgrown trees and brush, it was surrounded by flowers, small evergreens, and planted trees. Above the entrance was a huge golden cross with the words "Oley Washington Sr. Youth Center." Several huge windows were in the front of the building.

As Haywood and Lilly sat in their car admiring the new building, a helicopter flew overhead. They both stepped out of the car and looked up, watching the helicopter land in a small space between the woods and the new parking lot.

"Sweetheart! It's a military helicopter, and I'll bet it's the colonel and his mother," Haywood shouted over the noise.

As the helicopter doors opened, Lilly was shocked when she saw her college roommate, Lori Docker. Now she understood why she couldn't get a hold of Lori and tell her about the youth center grand opening. Lori walked down the helicopter ramp holding hands with the colonel and his mother. Haywood looked at Lilly, nodding his head and laughing. The colonel, his mother, and Lori walked over to the Runyans' car, where Lori hugged Lilly and Haywood. Then Oley and his mother hugged them. Father Pario came rushing over, shouting, "It's a miracle! It's a miracle!"

Lilly took a deep breath. "Lori, I didn't know you knew the colonel."

Lori smiled and hugged Colonel Oley Washington Jr. "We've known each other since Oley contacted me."

Lori laughed, and Colonel Washington explained, "You told me about her, so I decided to give her a call." Lilly nodded. "But I never gave you Lori's phone number." Lori and the colonel laughed.

"Oley called the children's hospital and requested to talk to me," Lori explained. "He invited me to come to Nevada and meet his mother. Since you told me about him, I wanted to meet him. You were right. Oley is everything you told me and more. We've been seeing each other whenever we can."

Lilly still looked surprised. "Did you fly in the helicopter from Nevada to here?"

The colonel shook his head. "We picked her up at her hospital heliport and brought her here with us."

Father Pario suggested they enter the new youth center, but when Haywood saw several government vehicles pull into the parking lot, he stopped. "Let's wait a moment," he said. "Agents Ted Ruffen, Bullord, and Bicker are here."

The agents joined them, and after several handshakes, they all walked into the new youth center. Sister Martha and Father Steven greeted them. "Haywood, did you get your new book published yet?" Sister Martha asked.

"I'm still working on it," Haywood said.

Fred Roll, with his employees Sally and Barney, were bringing food and drinks to the people celebrating the new youth center. Father Pario stepped to the podium and requested that Cardinal Blessings come to the podium and offer a prayer. Next, Father Pario invited Colonel Oley Washington Jr. to join him at the podium.

Colonel Oley Washington Jr. walked up to the podium and hugged Father Pario before speaking.

"My mother and I want to thank everyone here, the builders who built this new youth center, and all the volunteers who worked on this project." Then he requested that Haywood and Lilly to join him. Holding hands, they walked up to the podium. The colonel was tearing up as he told about his new life that had been created because of Haywood and Lilly. "If it wasn't for these two people, I would still be getting my food out of dumpsters and living in a bank safe. Haywood, please tell the people here what you and Lilly did to rescue me."

Haywood again hugged Oley. "Lilly and I were heading home from Canada when she said to me on a bus, 'Haywood, do you believe people living in these neighborhoods are living like us?' I told her, 'Honey, you just gave me an idea to write a book about someone living in this neighborhood who is just like us.' Father Pario was responsible for me finding Oley Washington Jr."

Oley replied, "Have you started writing your book yet?"

Then all the adults sitting in the youth center stood, clapping and hollering, "We want your book!"

When things quieted down again, Oley continued his speech: "Lilly, you helped me in Washington, DC, when you called the ambulance after I was beaten up by those college kids. The day I saw you again, my memory returned. My mother and I can't thank either of you enough for what you have done."

Father Pario pointed at Haywood and told the crowd, "Because of all these wonderful people, we now have a Oley Washington Sr. Youth Center to be used by everyone. This is a miracle, a miracle by our God."

Everyone stood up, clapping.

Later Haywood had an opportunity to talk to Colonel Washington alone. "Colonel, it's working. When the formula pellets are put into a solid steel container, they mass into a blob and continue to reproduce more pellets."

Agent Ted Ruffen approached Haywood and Oley, telling them, "I heard you talking about the pellets. You must be very careful using them, as the pellets, known as Can Do, now belong to our military and are classified as top secret. Don't do anything with the pellets that could get you arrested."

His face turning red in anger, Haywood asked, "Are you planning to arrest me if I continue to use the pellets?"

Ted nodded. "We've always been great friends, Haywood," he whispered. "Just be careful when using them."

"Are you using any pellets?" Haywood asked Ted.

Ted laughed. "My gas tank is full of water and pellets."

Oley broke in, "You both need to be very careful so no one sees you using the pellets. My father would be happy hearing that some of the public are using the pellets."

As Haywood and the colonel returned to the youth center, they found Lilly, Sister Martha, and Lori talking to each other. Lori gave the colonel a hug and said, "Is everything okay?"

The colonel smiled. "Yes. Haywood and I had much to talk about."

"So did we," Lori said. "Can we see the new swimming pool before we leave?"

They walked out on the swimming pool deck, watching families with kids swimming. Father Pario approached them, asking, "Isn't this a wonderful place that God, Haywood, and Colonel Washington have given to our neighborhood? Thank you, thank you. It's a miracle. It's a miracle."

They left Father Pario at the pool and approached the military helicopter. People who had been attending the youth center celebration were now taking pictures there. Lori, Lilly, Haywood, the colonel, and his mother shared more hugs and said good-bye and God bless. Haywood watched Lori, the colonel, and his mother enter the helicopter and leave the area. Haywood and Lilly held hands walking back to their car. Lilly, smiling, looked at Haywood, saying, "Honey, I am so proud of you. This neighborhood will always remember us for what you did. We have such great new friends in Lori, Colonel Washington, and his mother. I'm so glad we could be a part of this accomplishment to give Oley's father such a great and lasting memorial."

Haywood responded by hugging and kissing Lilly.

They returned home and found their home untouched, the new furniture looking as nice as the day they purchased it. Haywood checked the blob of pellets still in the steel case and still producing. Lilly looked at Haywood, smiling, realizing she was not going to stop Haywood from using the pellets.

www.ingramcontent.com/pod-product-compliance
Lightning Source LLC
Chambersburg PA
CBHW031908200326
41597CB00012B/547